SOLAR ENERGY

The Wykeham Science Series

General Editors:

PROFESSOR SIR NEVILL MOTT, F.R.S.

Emeritus Cavendish Professor of Physics
University of Cambridge

G. R. NOAKES

Formerly Senior Physics Master
Uppingham School

The Author:

JOHN I. B. WILSON obtained his Ph.D. in applied physics and electronics at Durham University, in co-operation with IRD Co. Ltd, Newcastle-Upon-Tyne, with a thesis entitled *Thin Films of CdS and the CdS/Cu₂S Solar Cell*. After a period at St Andrew's University in the Wolfson Institute of Luminescence, he was in 1975 appointed Wolfson Research Fellow in the Department of Physics, Heriot-Watt University, to establish a group developing Schottky-barrier solar cells. The work is now concentrated on amorphous silicon solar cells.

The Teacher:

H. G. BROWN, Rutland Sixth Form College, Oakham, Leicestershire.

SOLAR ENERGY

JOHN I. B. WILSON
Heriot-Watt University Edinburgh

WYKEHAM PUBLICATIONS (LONDON) LTD
(A member of the Taylor & Francis Group)
CRANE, RUSSAK & COMPANY, INC., NEW YORK
1979

Sole Distributors for the U.S.A. and Canada
CRANE, RUSSAK AND COMPANY, INC., NEW YORK

First published 1979 by Wykeham Publications (London) Ltd.

Library of Congress Cataloging in Publication Data

Wilson, John I B
    Solar energy.
    (The Wykeham science series; 56)
    Bibliography: p.
    Includes index
    1. Solar energy. I. Title

TJ810.W56   621.47   79–19525

ISBN 0–8448–1331–1

Printed in Great Britain by Taylor & Francis (Printers) Ltd,
Rankine Road, Basingstoke, Hants RG24 0PR.

Preface

The importance of the Sun for man's existence is largely ignored now that we can supply our own heating and lighting, and are not responsible for the direct production of our own food and clothing. It is true that most people are glad to welcome the Sun after a long cold winter, but its annual 'return' no longer attracts the worship accorded to it in former times, nor are the predictions of eclipses a useful way of instilling awe into a populace. However, life does rely on the continuance of solar radiation, and our controlled energy supplies are still largely solar-derived. It is not impossible that we might have to return to a greater consideration of solar radiation, perhaps necessitating a change in lifestyle and social structure, if nuclear fission is rejected as an energy source, and nuclear fusion is unworkable. Solar energy scientists are not treated with as much cynicism as they were by Swift in his *Gulliver's Travels* ('extracting sunbeams out of cucumbers . . . to warm the air in raw, inclement summers'), for much of the technology for converting solar radiation to alternative energy forms now exists. With an improvement in the economics of solar radiation conversion, together with better energy-storage schemes, solar energy could again provide the world's requirements as it once did before.

John I. B. Wilson

Acknowledgements

Many companies and individuals have responded promptly to my request for information on their own particular aspect of solar energy. In particular, I wish to thank the following: J. H. Apfel of Optical Coating Laboratory Inc., M. D. Archer for information on photochemistry, J. P. Coffey for information concerning his solar still, R. H. Douglass of TRW Inc., A. J. Gorski for data from the Argonne National Laboratory, J. M. Heldack of Varian Associates, R. J. Mytton and IRD Co. Ltd, D. G. Schueler of Sandia Laboratories Spectrolab Inc., and Ferranti Ltd, who have all give information concerning their solar cells, P. G. Smith of Kent Solartraps Ltd, and the UK section of the International Solar Energy Society. W. J. Firth of my own Department has provided a great deal of information on photosynthesis, storage and economics. I am grateful to G. R. Noakes for indispensable advice on both content and style during preparation of the manuscript, which was painstakingly typed by Janice MacLeod.

Contents

1. Energy

This text is principally a discussion of the physics involved in applying solar energy, but it is not possible to examine the subject in isolation from other energy resources. We shall find that it is difficult to separate any discussion of the *technology* of novel energy resources from *economic* and *social* considerations, for we already possess several alternative techniques which could be applied now if it were not for these other limitations. It must also be said that a study of 'energy' alone would be based on an ill-defined concept. There is no simple description of what energy *is*, as opposed to a description of its effects.

Another problem arises when we speak of energy *consumption*, for is it not true that energy is conserved? Although energy can apparently be *produced* in nuclear reactions by converting some mass into energy, it is generally neither gained nor lost in any process, only converted into another energy form. It is not always obvious that energy in its many guises is the same 'thing', but any energy unit can be applied to any energy form. Once we used kilowatt-hours (kWh) for electrical energy, British thermal units (BThU) for heat energy, and calories for thermal and chemical energy and so on, but now it is becoming common to use the joule (J) for all forms. When energy is used to do work it is successively reduced to low-grade heat, i.e. small-scale molecular motion. The energy has then become unavailable for normal purposes, and has undergone an increase in *entropy*, or disorder. Strictly there is no 'renewable' source of energy, but solar radiation is often called 'renewable' since the Sun's lifetime is so long on a human scale. In contrast, fossil fuels are not renewable on the scale of present consumption, and production by photosynthesis.

1.1. Energy resources

We have available a wide range of possible energy sources, each capable of doing the same useful work for us, but each has special characteristics which make it particularly suited for use only in certain ways. The fossil fuels, coal, oil and gas, are not only useful for producing thermal energy but also for their chemical constituents. Although these fuels are a convenient form of concentrated solar energy, it is possible that their future importance will be as valuable chemical feedstock. There will always be plenty of energy available in the world, but not necessarily in a form suitable for a modern industrialized nation.

The resources available are in many states. These can, at present, be identified as:

(a) Wood, plant wastes, animal wastes.
(b) Coal, oil, gas, peat, oil shales, tar sands.
(c) Solar radiation, hydraulic, wind, waves.
(d) Tidal.
(e) Geothermal.
(f) Nuclear fission.
(g) Nuclear fusion.
(h) Chemical.

There has been a tradition of underestimating the size of the world's fuel reserves, but consumption has risen to such a level (and is increasing at such a rate), that now we can make good guesses of remaining reserves and their lifetime without great error in estimating the time left before they are exhausted. It is helpful to have some idea of how much energy is probably still available from these resources.

(a) *Wood and wastes:* These are of great importance in the economy of many countries, although total world consumption may appear to be small; since $\sim 0.1\%$ of the solar radiation falling on the Earth's surface is converted to vegetable matter by photosynthesis there is a great store of energy, but obviously not all of this $\sim 3 \times 10^{21}$ J can be used.

(b) *All fossil fuels:* Total recoverable reserves $\sim 3 \times 10^{22}$ J.

(c) *Solar energy:* Total received at the Earth's surface $\sim 4 \times 10^{24}$ J/year; this is converted to other energy forms which may be used instead of directly applying solar radiation:

(i) hydraulic: $\sim 35 \times 10^{18}$ J/year, of which less than 50% is available most of the year.

(ii) wind: Potential generation using *all* winds would be equivalent to less than 1% of the solar radiation received, say $\sim 6 \times 10^{21}$ J/year.

(iii) Ocean thermal gradients: theoretically available 6×10^{18} J/year—but many suitable sites are far from land.

(iv) Waves: this resource comes from the Sun via wind and gravitational energy and is very difficult to assess: only sites near to land are thought to be useful.

(d) *Tidal:* This energy derives from the Moon's gravitational pull and the total world dissipation of tidal energy is $\sim 10^{20}$ J/year, of which we might use 10^{18} J/year.

(e) *Geothermal:* Natural geothermal heating from radioactive decay, down to 10 km, is $\sim 4 \times 10^{22}$ J, of which we might use 10^{18} J/year.

(f) *Nuclear fission:* There is some difficulty in estimating the known reserves of uranium and thorium ores, and in addition to this they are strongly dependent on the prices we are prepared to pay for their extraction; say about $6 \cdot 7 \times 10^{22}$ J are realizable.

(g) *Nuclear fusion:* Deuterium and tritium are effectively in infinite supply if they can be extracted from the oceans: $> 10^{28}$ J.

Since present world consumption is somewhat less than 3×10^{20} J/year, but is increasing, it appears that no single alternative technology suffices to provide our requirements, although a combination of several with greater use of solar energy is attractive. It is important to realize that most of these alternative sources are available *each year* whereas fossil fuels alone would be exhausted in about 160 years at our present rate of consumption. Fusion is the only energy source with the capability of providing a large amount of energy for an 'unlimited' period, and we have yet to demonstrate *controlled* nuclear fusion. Most energy sources have environmental hazards attached, but the most extreme are from nuclear fission, thus many people are unwilling to rely totally on a nuclear fission economy.

1.2. *Energy consumption*

The non-uniform distribution of mineral resources has led to recent crises in conventional fuel supplies. At present Africa consumes less energy than Europe, and has about 350 years supply of its own fossil fuels at the present rate of consumption, whereas Europe has only about 54 years supply at its present and predicted rate of consumption. As these fuels become scarcer the pattern of use will change, and the price will go up. Lower grades of these fuels will then be economically worth producing.

Although we shall be considering the use of solar energy as part of the answer to the problem, we must realize that there will be an equally important shortfall in many other mineral resources unless conservation and recycling are used on a large scale. As with energy, we have the chance of supplying the demand as long as the materials are used sensibly. Current economically recoverable reserves of some important metals and semiconductors are given in table 1.1, together with their relative abundance in the Earth's crust. Given *unlimited* energy it would always be possible to extract these from less and less concentrated ores, and from the oceans.

TABLE 1.1. *Abundance and exploitation of certain common elements.*

Element	Abundance in the Earth's crust (p.p.m.)	Economically recoverable (10^6 g)	World annual production (10^6 g)
Si	276 000	$\gg$	$> 3 \times 10^3$
Al	80 000	3×10^{12}	6×10^7
Fe	50 000	2×10^{12}	4×10^8
Cu	64	3×10^8	5×10^6
Ga	18	1×10^5	1
As	2	2×10^7	5×10^4

10^6 g = 1 tonne. p.p.m. = parts per million.

Most available figures for energy consumption concern the USA, as one of the major users, but some figures are universally applicable. These are for the energy consumed during the production of various materials. The idea of accounting for all the energy needed to produce, say, a tonne of steel can be useful, but it is always difficult to decide how far back in the chain it is necessary to go. For instance, should the food consumption of the workers be included? Table 1.2 shows how false savings with respect to the real cost of production

TABLE 1.2. *Energy cost for manufacture of various materials.*

	GJ tonne^{-1}
Plastics (powder)	100
Steel (bar)	48
Al (ingot)	330
Cu (bar)	60
Glass	30
Portland cement	8

can be encouraged by the use of 'cheap' aluminium and plastics instead of steel and glass. Despite the differences in density, and in machining/forming costs, plastics are still very energy-expensive compared with older materials.

It is equally important to use the correct form of energy for any particular application, although once again there is a short-term incentive to use the 'cheapest' form rather than the most suitable. Thus for heating purposes it is not overall efficient to use electricity however convenient this may be. Electrical heating appliances are only about 20–30% efficient compared with gas-burning heaters which are about 40–70% efficient, when the energy consumed in production and transmission has been included. In themselves electrical appliances are efficient converters of energy (see table 1.3),

TABLE 1.3.   *Conversion efficiency of some appliances and machines*

Light bulb	5%
Internal combustion engine	25%
Fluorescent lamp	20%
Nuclear reactor	39%
Steam turbine	46%
Electrical storage cell	72%
Dry cell	90%
Hydraulic turbine	92%
Electric motor	93%
Electric generator	98%

but the initial generation of electricity and its inefficient transmission down resistive power lines mean that less than one third of the energy is usefully converted. Much of the waste heat produced at the power stations could be usefully employed for district heating, as it is in a few places, but the energy loss by Joule (I^2R) heating in the power cables can only be reduced by using superconducting materials cooled by, say, liquid helium—probably not energetically favourable. This waste of prime energy sources extends over the whole range of applications—industrial, commercial, domestic, and particularly in transport. Since about a quarter of the annual energy consumption in the UK is used for transport, and since this is often less than 25% efficient in its power unit, we can see here a tremendous annual loss of high-grade fuel. Over all applications the conversion of energy into useful work is probably around 50% and the rest is degraded to useless heat.

In contrast to most modern developments, agriculture is tending to reduce its energy consumption. By the use of weed killer such as paraquat, and direct drilling of seed rather than ploughing, harrowing and so on before sowing, there is less energy used to grow the crops despite the energy consumed in the production of weed killer. This technique is not necessarily environmentally bad for it is said to improve the structure of over-ploughed soils.

The problem that we are faced with is really one of *growth* in population and consumption. So far, the industrialized nations have relied on coal and then oil for their rapid advance, and the oil supply, in particular, is reaching the end of its lifetime. The UK now uses 45% oil, 35% coal, 16% gas and 4% nuclear and hydroelectric in its annual energy consumption. Certainly coal can continue to be used for several decades yet, and if it is burnt in central power stations the pollution from the combustion by-products can be more easily kept under control than was once the case. This still leaves the question of how to transmit the energy efficiently to the user, and despite cable losses, electricity is probably the most convenient form. An alternative has been suggested: hydrogen is a clean fuel with high calorific value and despite the handling hazards there are many supporters of this proposal. It would be a convenient fuel for production by off-shore power plants which might be nuclear-powered, wave-powered, or by ocean thermal energy-converters. Since different nations have different priorities for their future development there is no one answer. It is also by no means certain that the development *must* imply greatly increased energy consumption per capita, although some correlation between gross national product and energy consumption has been shown. If there is to be growth in production of a nation's industry, and this appears to be the only way that many countries can maintain social equilibrium, then it will have to use nuclear fusion, fission or a combination of very large-scale alternative energy projects.

The question of whether the UK should rely on a nuclear breeder reactor for its future power is being hotly debated at present. Certainly there is a limit, as we have seen, to the expansion of non-breeder reactors, but an intermediate stage using enriched fuel may avoid the worrying build-up of plutonium stockpiles and waste, with the dangers of leakage or terrorist activity. This is being considered in many countries and concerns more than just the technical problems. Even if decisions are taken soon there may be a short-term problem in the UK caused by delay in building sufficient new power stations for use when present fuels become scarce. The longer term solution must be a grandiose solar collecting scheme or fusion (if it can be controlled) combined with greater use of geothermal energy.

Let us look at the *basic* energy consumption of a single person, and compare this with *actual* consumption in an industrialized society. This will give some idea of the population which could be supported by the Earth, if energy was the only limitation. The food needed by an adult human obviously depends on the climate and on the occupation of the person, but a rough estimate is 10–20 MJ/day. In the USA each person uses something like 860 MJ/day! Thus a World population of about 10^{10} should be supportable on energy grounds alone; present population is about 4×10^9. However, we are again dealing not only with a scientific or agricultural problem, but also with a political/social question. Certainly this population could be provided with food, but not under inefficient Western practices of consuming much animal protein. Although there are very large differences in the farming efficiency, climate, incidence of pests and diseases, and eventual waste, around the World, a population of 10^{10} *could* be fed from photosynthesized carbohydrate. If we take the total land area available, subtract an allowance for each person's living requirement, and assume an *overall* efficiency for sunlight into delivered food of 0.1% for the remaining land surface, then this is found to be about the maximum stable population. The supply of other human needs (especially water) would provide similar checks on unlimited population growth. It appears that the social consequences of expanding the population greatly overshadow the limitations imposed by energy requirements alone.

1.3. *Energy conversion*

The successive conversion of energy into other forms and ultimately into heat is the only way in which we are able to consume energy. The study of the transfer of heat and its application to work is based on the laws of thermodynamics. We shall make use of some thermodynamic reasoning in order to understand the limiting efficiency for the conversion of solar radiation into other energy forms. For example, it is basic to thermodynamics that no engine can be more efficient than one working on a Carnot cycle between the same source of heat and sink for rejected heat. The Carnot cycle is an ideal reversible cycle in which heat is transferred isothermally (both at source and at sink), and in which there are no thermal or frictional losses (which cause all real engines to be irreversible). Even practical heat-engine cycles are usually idealized to the extent that a process is supposed to take place under, say, isobaric conditions whereas in reality there would be a small but significant pressure change.

Many thermodynamic cycles are in use, or have been proposed. They may be represented on change of state diagrams (showing pressure, P, volume, V, and temperature, T) to show how the working fluid passes around a cycle taking up energy, doing work, and rejecting waste heat (see *Appendix A*). The Stirling cycle (for a gas) and the Rankine cycle (for a vapour) are of interest for solar energy converters. Rankine cycles are sometimes *open* and so are irreversible, which reduces the maximum attainable efficiency. The Stirling cycle is still not widely used, although an open version is part of gas liquefying plants in which work done by the gas helps to cool it down. Closed-cycle Stirling engine designs based on conventional

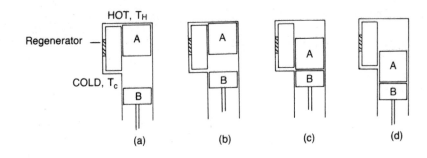

Figure 1.1. Closed-cycle Stirling engine with a single displacer piston (A), and a power piston (B). (*a*) Gas in cold chamber, displaced there by A; (*b*) gas compressed at temperature T_C; (*c*) gas displaced into hot chamber (T_H); (*d*) gas expanded, pushing both pistons down.

heat engines must overcome the related problems of making a gas-tight seal around the pistons whilst reducing the frictional losses at their slow running speed. (The cycle is potentially capable of conversion efficiencies up to $\sim 70\%$ of the Carnot efficiency for the same heat source and sink, compared with less than 60% of the Carnot efficiency for a Rankine cycle.) Two pistons are generally used (see figure 1.1), one being a floating displacer piston which transfers the gas from the hot zone of the cylinder to the cold zone, and back again. It is important that the displacer and working pistons move with the correct phasing so that the expansion of the heated gas drives the working piston effectively. An interesting design produced at AERE Harwell, called the 'Fluidyne' heat engine, uses a liquid piston to overcome the sealing and friction problems. This is a simple machine, with few mechanical moving parts, which produces a slowly oscillating liquid column ($\sim$1 Hz), and which could be suitable for solar-powered water pumping. Its low efficiency ($<$1%) is not a drawback when compared with the simplicity of construction and low operating costs.

1.4. *Conclusion*

Our dwindling reserves of conventional fuels will force us to look elsewhere for future energy supplies. 'Alternative technologies' using such renewable resources as Sun, wind and waves are capable of providing a significant part of this energy, but will require novel ideas for the efficient conversion and transfer of energy to the consumer. Nonetheless, the basic physical laws which give us an understanding of present-day energy conversion will continue to be applicable.

2. The Sun

2.1. *The Sun as an energy source*

The Sun is the source of nearly all our energy (with the exception of radioactive sources and the tides), and will continue to be the most important unless controlled nuclear fusion and fission reactors are to be used. Many people would prefer our nuclear reactor to be sited $1 \cdot 5 \times 10^8$ km away, the distance of the Sun from the Earth, rather than place increasing emphasis on various fission reactor designs. If man is to use yet more energy than at present, there will be difficulties in obtaining enough from the solar radiation reaching the Earth. Even including the resources of hydro-power and agriculture which are solar derived, it will become increasingly difficult to run an industrial society from solar radiation alone. Yet the Earth receives from the Sun 5000 Q/year,* which is far greater than the present world energy consumption of $0 \cdot 25$ Q, of which only $0 \cdot 1$ Q is derived from solar 'income' (hydro-power and green plants) rather than from solar 'capital' (fossil fuels). To see why solar radiation is not efficiently used by man, we must realize that wave, wind and hydraulic power are all solar driven.

If the Earth is intercepting 5×10^{24} J/year from the Sun, this means that the annual emission from the Sun is 10^{34} J, and this has probably been so for 10^{10} years. The total energy loss from the Sun (10^{44} J) is most likely to have come from nuclear *fusion*, for the Sun is known to consist mainly of the light elements, hydrogen and helium, and does not contain a large amount of the heavy elements which can undergo *fission*. Furthermore, the high temperature within the Sun ($1 \cdot 5 \times 10^7$ K) is consistent with the thermal energies required by fusion. There are two likely sequences for the conversion of hydrogen into helium within the Sun: a proton–proton reaction, and the carbon cycle, but the end result is the same. The reaction of four protons ($^1_1$H

*1 Q is a convenient measure for comparing large amounts of energy:

$$1 \text{ Q} \simeq 10^{21} \text{ J} \simeq 3 \times 10^{14} \text{ kWh.}$$

nuclei) to one $^4_2$He nucleus is accompanied by a 0·7% loss of mass, which appears as energy.

Some of the energy produced is taken from the Sun by escaping neutrinos, but most appears as radiant energy. In fact, the power density is not large enough for these processes to be worth copying on Earth, and the deuterium-based fusion plants proposed will give higher energy densities, albeit at a higher threshold temperature. The Sun contains sufficient hydrogen that its conversion into helium *does* give the required total energy of 10^{44} J. It is estimated that over 5×10^{16} kg hydrogen are converted into helium and energy each day.

2.2. *Solar radiation*

The Sun is very nearly a black-body emitter with an effective surface temperature of about 5800 K, which means that the spectral distribution of the radiation received by the Earth is broad (see figure 2.1). Impinging on the Earth's atmosphere is a total of 1·3 kW m^{-2}, which is known as the solar constant, 95% of which is at wavelengths shorter than 2 μm. Much of this radiation is scattered back into space, and a further large proportion is selectively absorbed by the various gases in the atmosphere, with the result that after a single vertical

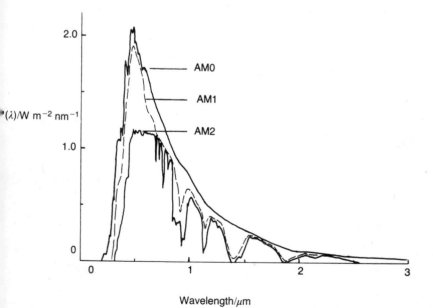

Figure 2.1. Solar spectral irradiance ($P(\lambda)$) for different air masses.

transit through the atmospheric layer there is only about 1 kW m^{-2} reaching us on the surface. Figure 2.1 also shows the selective filtering action of various thicknesses of atmosphere (i.e. oblique transit paths). The 'air mass' is determined by the thickness of atmosphere in the radiation path, using the cosecant of the Sun's altitude. In Scotland the low altitude of the Sun in winter provides less than air mass two (AM2) irradiance. The AM1 solar spectrum, most conveniently used as a standard, may be roughly divided into 51% infra-red, 40% visible and 9% ultraviolet. At higher air mass numbers, the selective absorption of atmospheric constituents in the infra-red waveband is even more prominent.

Figure 2.2 outlines the Earth's radiation balance with the Sun. The

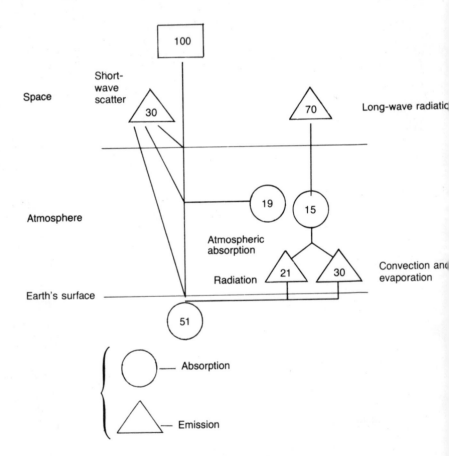

Figure 2.2. Radiation balance of the Earth.

solar input is balanced by a combination of short- and long-wavelength radiation returned from the atmosphere and the Earth's surface. The spectral distribution of this emission as seen from space is also closely that of a black body—with an effective temperature of 252 K. We shall later be discussing the importance of radiant emission to the Earth from clouds, and shall find that this has an appreciable warming effect.

Solar energy is the only source from which we can use more energy than at present, without contributing a new thermal energy input to the atmosphere. However, it should be realized that although the *total* thermal input may not be changed, its *distribution* may be. Urban energy consumption is greater than the solar input to the area, and agricultural consumption of energy is less than the local solar input. (A city like New York feeds around 600 W m^{-2} of waste heat into the atmosphere, compared with the average net natural radiant emission of less than 100 W m^{-2}.) Any large-scale use of solar energy for urban areas is likely to produce a change in local albedo of both collector site and user site, unless the reflectivity of the surroundings is altered to compensate for the changes produced. Small-scale applications, the most important for some time to come, should have no thermal pollution side-effects.

Measurements of the various components of solar radiation and of radiation balance at the Earth's surface, are being made at numerous sites as well as by orbiting satellites. This has demanded improved instruments and international standardization so that the different measurements can be compared with confidence. All measurements are referred to the same standard atmosphere. Most of the available data on solar radiation at the Earth's surface are for insolation on a horizontal plane, rather than for the tilted plane used by many solar collectors. A simple geometrical correction may be applied, but it is not easy to allow for reflections from the ground or nearby objects, nor is it true that the diffuse (scattered) and direct components of solar radiation are in the same ratio for different orientations.

The most commonly available data are of global irradiance, measured by a pyranometer (solarimeter): this is the irradiance received by a horizontal surface from both Sun and sky. Less common is the subdivision of this into direct (or beam) solar irradiance, measured by a pyrheliometer, and diffuse (or scattered) irradiance, measured by a pyranometer screened by a ring from the path of the direct sunlight. The spectral distribution of solar energy is of great importance for quantum-type converters like solar cells, but is more difficult to determine absolutely. Either a monochromator (with prism or diffraction grating) or a set of narrow-band-pass filters, combined with a calibrated detector is suitable. Both pyranometers and pyrheliometers may use a thermocouple detector arrangement to sense the temperature difference between adjacent black- and white-coated metal surfaces, covered by glass domes to eliminate atmospheric contamination. The surfaces which intercept the solar radiation should be perfectly matt and should withstand prolonged exposure to the Sun. Pyrheliometers use a tracking mechanism to keep them pointing at the Sun—only a small area of the sky around the Sun is viewed by the sensors, through a tube with optical stops. All instruments require *frequent* calibration checks, cleaning and maintenance.

Not only does the solar radiation received vary both with hour and season, but it arrives with a low energy-density. These features make industrial use of solar radiation difficult, even if we ignore the fluctuations caused by cloud, unless some efficient form of energy storage can be devised. Until comparatively recent times man relied exclusively on the energy storage provided by plant-life to provide a 'condensed' form of solar energy. Before the mid nineteenth century, man lived on the solar input stored in living plants (e.g. wood), but he has become increasingly reliant on the longer-term storage provided by fossil plants or animals (e.g. coal, oil, gas). Today we still have the problem of inexpensively collecting a sufficient quantity of solar radiation for *any* application. Since at most we can only expect an input of 1 kW m^{-2}, a chain of collection, conversion, and final application with a modest overall efficiency of 10% may need 10 m^2 collector for even the smallest application. To generate electrical energy on the scale of a present-day large conventional power station (1000 MW) would require a 10% efficient solar collector/converter to be half the area of Wales (10^{10} m^2), and this should cost less than £5 m^{-2} of collector area to compare with the capital cost of a fossil fuel plant.

What is required is a cheap collector of solar radiation which can then deliver it to an energy converter, in the same manner that water evaporated from the sea by the Sun's radiation might fall on mountains to be collected into a reservoir where it has stored potential energy available for conversion into electrical energy. On a smaller scale it may be economically worth while to collect solar radiation by a cheap optical system of mirrors or lenses for delivery to a smaller expensive converter. This is only possible if the sun is directly visible, for the diffuse radiation which makes the sky bright on overcast days cannot be focused to an image. Even on a clear day the diffuse radiation can amount to 15–20% of the direct solar radiation. Optical concentrators are of little use in the UK over a whole year, and we must make do with large-area collectors of a different type.

Averaged over an entire, year, the UK mean solar radiation is 125 W m^{-2}, which is only 40% of that received in more favoured places, and of this input 60% is from scattered radiation. Since the spectral distribution of scattered sunlight is different from that of direct sunlight (figure 2.3), a solar collector which is wavelength-dependent will have different efficiencies for different atmospheric

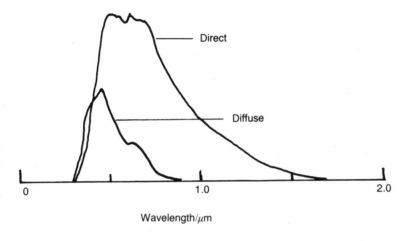

Figure 2.3. Spectral distribution of direct and scattered sunlight (AM2). (Reproduced by kind permission of the UK section of the International Solar Energy Society.)

conditions. Despite the apparently low annual solar irradiance received by the UK compared with other countries (figure 2.4), the main difference is in the winter irradiance. Detailed radiation maps for the UK are available from the Meteorological Office, but these are extrapolated between the widely spaced observation stations and so no accurate calculation of the irradiance for any particular site can be made. It is obvious that in heavily industrialized areas there will be more atmospheric scattering of sunlight than in clearer rural districts, and although Scotland has lower insolation in winter than, say, London, it has clearer skies and longer hours of daylight which partially compensate for the loss. The atmospheric turbidity, the allowance for the scattering effect of dust and water droplets, is one

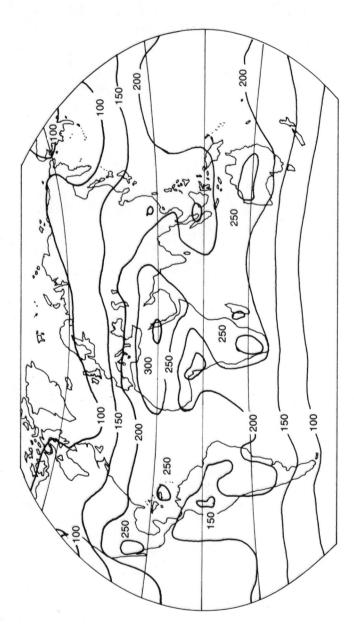

Figure 2.4. Annual mean global irradiance on a horizontal plane. (Isoflux contours are in W m⁻².) (Reproduced by kind permission of the UK section of the International Solar Energy Society.)

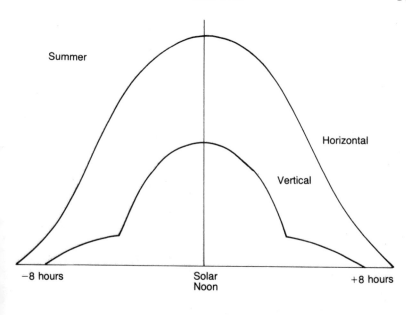

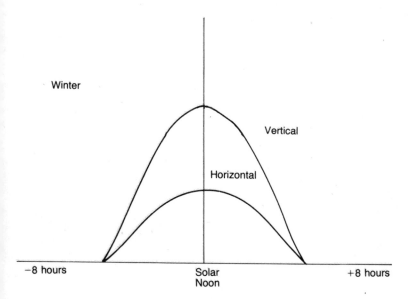

Figure 2.5. Typical hourly variation in total irradiance on a horizontal or vertical (south-facing) plane.

of the more difficult parameters to assess, and it has a great effect on the amount of direct sunlight.

One aim of collecting solar radiation data would be to afford predictions of usable solar power for any chosen site merely from a knowledge of its latitude and longitude and the type of geographical situation. One could then choose the best collector system without trial and error. At present this is only possible for a few favoured sites. Even the question of the best orientation for a fixed solar collector cannot be easily answered. Although it is common to suggest a tilt angle of 10° or 15° more than the local angle of latitude, this is not necessarily the best angle in the UK because of the large proportion of scattered radiation we receive. In winter, a south-facing *vertical* collector will allow for the low altitude of the Sun, when it is visible, but on overcast days the sunlight is scattered and comes from overhead, in which case a *horizontal* collector would have the advantage. Figure 2.5 shows the typical hourly variation in the total insolation on horizontal and vertical surfaces.

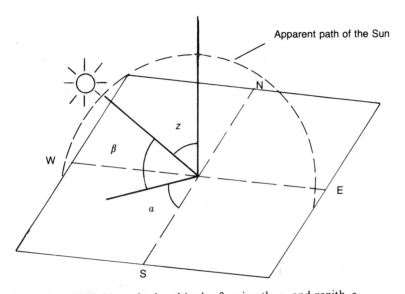

Figure 2.6. Definition of solar altitude, β, azimuth, a, and zenith, z.

β = angle between line from Sun to Earth's centre, and tangent to Earth's surface.

a = angle between south meridian, measured in horizontal plane westwards, and the direction of the Sun.

z = angle between the line from the Sun to the Earth's centre, and the normal to the Earth's surface.

A method of calculating the hours of daylight for any location would be useful, and if coupled with a knowledge of the Sun's apparent path across the sky would allow theoretical comparisons to be made of the relative performance of fixed and tracking collectors, as well as setting an upper limit to the amount of collected energy. This it is possible to do, for the angle of the Sun above the plane of the celestial equator (the solar declination) is tabulated for various latitudes throughout the year. It is obviously only a geometrical problem, albeit a complicated three-dimensional one, to calculate the hours of daylight. These could be combined with the air mass number for the site throughout the year, and the 3% seasonal variation in the solar constant (due to the eccentricity of the Earth's orbit) to arrive at a figure for the maximum annual insolation for the site. The equations which give the Sun's position above the horizon are quoted below. The hour angle is the number of minutes from *solar noon* divided by 4 to convert to degrees. We should use solar noon as our reference because local 'clock' time is arbitrarily held constant in 24 time zones, each with a spread of degrees of longitude, whilst the Sun traverses a degree every 4 min. The altitude and azimuth angles are defined in figure 2.6.

$$\sin \beta = \cos L \cos D \cos H + \sin L \sin D \qquad (2.1)$$

$$\sin a = \frac{\cos D \sin H}{\cos \beta}, \qquad (2.2)$$

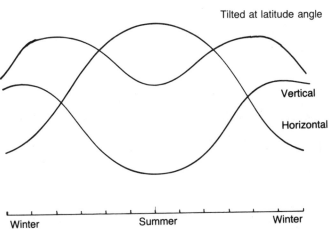

Figure 2.7. Seasonal variation in total irradiance on variously oriented planes.

where L = latitude, D = declination, and H = hour angle. These ang-les are also tabulated for several latitudes and certain dates.

Figure 2.7 shows the seasonal variation of the total insolation falling on a horizontal surface, a vertical south-facing surface and a south-facing surface tilted at an angle, L, to the horizontal. No account has been taken of cloud, which may be much more in evidence for the winter months in the UK, but nonetheless the advantage of a vertical collector in winter is shown, as is the additional advantage of a tilted collector in all but the summer months. In order to make some adjustment to this graph for diffuse radiation predominating in winter, a geometrical factor must be used (figure 2.8). *Direct* insolation on a tilted surface is equal to the

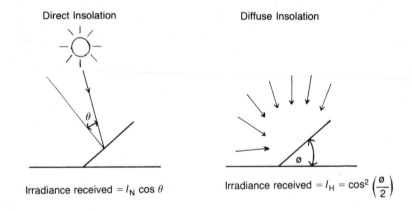

Direct Insolation

Diffuse Insolation

Irradiance received = $I_N \cos \theta$

Irradiance received = $I'_H = \cos^2 \left(\dfrac{\phi}{2} \right)$

Figure 2.8. Corrections to be made for insolation on a tilted plane, from data for a horizontal plane.

amount (I_N) received by the surface if it were normal to the Sun's rays, but multiplied by the cosine of the angle of incidence. *Diffuse* insolation on a tilted surface is given approximately by the amount received by a horizontal surface (I_H) multiplied by the square of the cosine of half the tilt angle. The diffuse component of the received radiation thus falls off slowly with increasing tilt angle, but may be made up to an extent not easily calculated by ground-scattered insolation and by reflections and radiation from nearby surfaces.

3. Light and matter

3.1. *Macroscopic behaviour*

The behaviour of materials is normally described in macroscopic terms. Here we are interested in the optical properties of solids and liquids, and so use such quantities as reflectance (ρ), transmittance (τ), and refractive index (n). If we wish to go further than just *describing* what happens when light and matter interact, then a more detailed understanding can only be gained from the *microscopic* behaviour of electrons, photons and similar entities. For instance, the three properties named above are wavelength-dependent and to see why this should be so, and especially to predict changes, the underlying physical processes must be examined.

We shall consider an ideal body which absorbs all radiant energy falling on it, and then describe the absorption of *real* substances by their deviation from this ideal. Since there would be no reflected radiation, the ideal body would always appear black. An ideal 'black body' absorber is also a perfect *emitter* of radiation. The wavelength (λ) distribution of the power emitted at temperature T is given by Planck's formula (3.1).

$$P(\lambda) = \frac{c_1}{\lambda^5[\exp(c_2/\lambda T) - 1]},\qquad (3.1)$$

where $c_1 = 3 \cdot 74 \times 10^{-16}$ W m^2, and $c_2 = 1 \cdot 44 \times 10^{-2}$ m K. The wavelength (λ_m) at which maximum emission occurs for any temperature is given by Wien's displacement law (3.2).

$$\lambda_m T = c_3,\qquad (3.2)$$

where $c_3 \sim 2900$ μm K.

A black body at room temperature emits most strongly near 10 μm. Figure 3.1 shows the black-body emission spectrum for various temperatures of particular interest for solar-energy conversion. A body at the temperature of the Sun's surface emits strongly at

SOLAR ENERGY

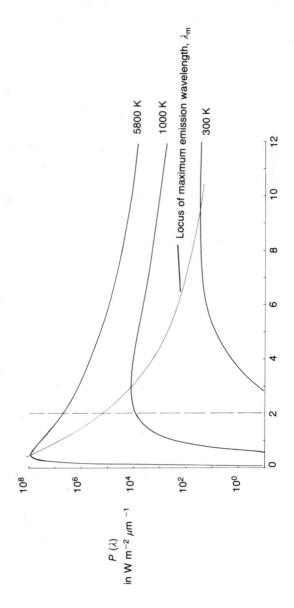

Figure 3.1. Black body energy distributions.

visible wavelengths, but for bodies at temperatures below 1000 K, practically all of the energy is emitted in the infra-red region of the spectrum.

The *total* power at all wavelengths, P, emitted per unit area for a particular temperature T is given by

$$P = \sigma T^4, \tag{3.3}$$

where σ is the Stefan-Boltzmann constant $= 5 \cdot 67 \times 10^{-8} \, \text{W m}^{-2} \, \text{K}^{-4}$.

A black body, then, absorbs radiant energy of all wavelengths, and re-emits energy with a wavelength distribution dependent *only* on its temperature. If it is in thermal equilibrium with a radiating source, then its absorptance (a) equals its emittance (ε), (both being equal to unity). This last statement is true of *any* body in thermal equilibrium, but in general $a = \varepsilon < 1$. We shall assume that $a = \varepsilon$ at any given wavelength, even if the surface is not in strict thermal equilibrium with the source (e.g. a solar collector and the Sun).

If a and ε are independent of wavelength but less than unity, the body is termed 'grey'. We shall be interested in surfaces which are one more step removed from the black body ideal: these have one value of $a = \varepsilon$ for the visible wavelengths, and another value for the infra-red, and are termed 'semi-grey'. The importance of such surfaces will become apparent when we consider the energy balance of solar collectors.

It is common to estimate the values of a and ε by reflection from the material in question, using a detector (and source) with only a *narrow cone* of acceptance (emission). This is similar to the spectrometer used in school laboratories. The value of a and ε thus found refer to *normal* (or nearly normal) *incidence*, rather than the *total* a and ε of the surface over the whole solid angle of a hemisphere. The total hemispherical a or ε (sometimes called the *global* a or ε) is the one usually required for solar energy studies but is somewhat more difficult to measure accurately.

To find a from reflectance data, one makes use of the energy relation between ρ, a, and τ

$$\rho + \tau + a = 1 \tag{3.4}$$

This is adequate for our present needs, but is a simplification, for it leaves out the question of radiation polarization. At present we are unable to make use of optical polarization effects in solar energy conversion, though this may perhaps turn out to be possible.

Figure 3.2 shows the reflectance of a glass surface for light plane-polarized parallel and perpendicular to the plane of incidence,

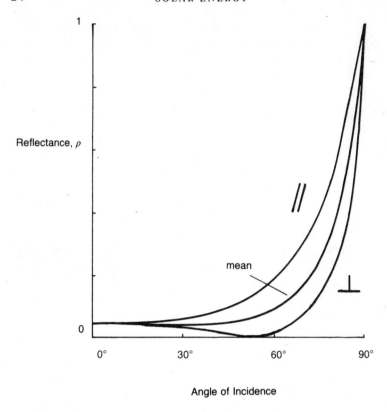

Angle of Incidence

Figure 3.2. Reflectance of a plane glass surface for light plane-polarized with the electric vector parallel to the surface ($\parallel$), perpendicular to the surface ($\perp$) and for natural light (mean).

as a function of the angle of incidence. For 'natural' light the average reflectance is given. *Scattered sunlight* is partially plane-polarized.

Even the concept of reflectance itself is not simple, for all surfaces may be somewhere between the extremes of a specular reflector (i.e. a mirror) and a perfectly diffuse reflector. This again emphasizes the importance of knowing the conditions under which an experimental figure was obtained.

Opaque materials will by definition have $\tau = 0$, and so $\rho = (1 - a) = (1 - \varepsilon)$. Materials are usually transparent to at least some parts of the full radiation spectrum.

Electromagnetic theory of normal incidence reflection from dielectrics gives Fresnel's equation relating the reflectance to the refractive indices of the dielectric (n_1) and the adjacent medium (n_2).

$$\rho = \frac{(n_1 - n_2)^2}{(n_1 + n_2)^2} \qquad (3.5)$$

This will be used in a later section when the subject of *reducing* reflectance from a surface is discussed. Complications arise when we wish to estimate the reflectance or transmittance of one or more *sheets* of glass. This is the situation found in flat-plate solar collectors with glass covers. The problem is that there are *multiple* reflections between the successive glass surfaces. Only $(1 - \rho)$ of the incident radiation passes from the first surface to the second (if $a = 0$), but some of this will be *returned* from the second surface. Figure 3.3

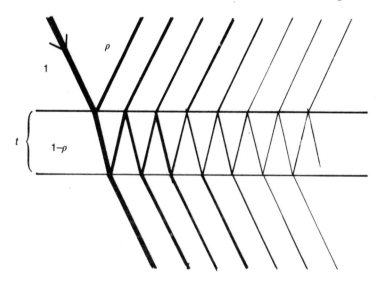

Figure 3.3. Multiple reflections from a glass plate in air.

makes this clearer. By adding all the components of radiation passing through the final surface of the i cover sheets, the total transmittance will be found.

$$T = \frac{(1 - \rho)}{1 + (2i - 1)\rho} \qquad (3.6)$$

We have assumed that the absorptance of these sheets is zero, but in practice even 'water white' glass, if thick enough, would eventually absorb all incident radiation. The usual rule for determining the transmittance of a thickness of material assumes exponential decay of the intensity as the radiation travels through the material, that is,

$$\text{if } \rho = 0, \tau = \exp(-\kappa d) \qquad (3.7)$$

where κ is called the extinction coefficient, and d is the path length in the material.

Let us return to the discussion which led to equation (3.6). By allowing the returned radiation from the *second* surface of a sheet of material to interact with radiation incident on the *first* surface, it is possible to reduce the reflectance of the first surface. The trick is to ensure that the *phase difference* between reflections from top and bottom surfaces is π. The simplest method of achieving this is to coat the materials with a single thin film having an *optical thickness* equal to a quarter of the wavelength, λ, at which ρ is to be zero. This means that the *geometrical thickness* is $\lambda/4n$. True zero reflectance is only obtained for normal incidence and at one value of λ unless a more complicated coating structure is used, but a *reduction* in ρ at other angles of incidence and other wavelengths will still be apparent (figure 3.4). Although reflectance is reduced, the radiation entering

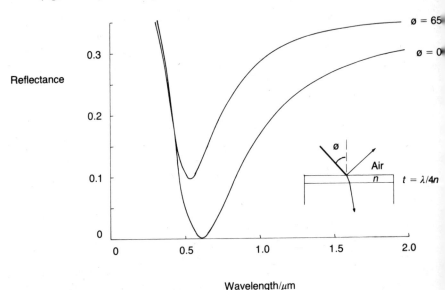

Figure 3.4. The effect of a single quarterwave anti-reflection coating on a surface.

the material beneath the film is still subject to *absorption*, and an anti-reflection coating cannot by itself produce 100% transmittance. With the cover plates of solar flat-plate collectors it is the value of τ rather than ρ which is important.

Throughout this discussion of the macroscopic behaviour of illuminated materials, we have used n for the refractive index. This is sufficient for ordinary transparent material, for simple solar collectors, but when semiconducting materials are involved it is often necessary to use the full complex refractive index, n', derived from Maxwell's equation for a conducting medium.

$$n' = n - \mathrm{i}\,k \tag{3.8}$$

where n, the 'real' part is related to the velocity of the radiation in the material, and is already familiar in our treatment of transparent materials, for which $k = 0$, and k, the 'imaginary' part is related to the absorptance, and is large for metals.

3.2. Microscopic behaviour

3.2.1. Metals

Metal layers form an important component of both the transparent and the absorbing parts of the solar collectors to be described. According to the free electron model of metals, they contain both free and bound electrons. Only the outer (valence) electrons of each metal atom are free to move through the bulk of the metal, and the remaining electrons are bound to the positively charged nuclei, which vibrate about a fixed position. We can immediately see two ways in which energy may be added to a metal: either to the free electrons or to the ions. The lowest energies which are absorbed are in the infra-red, and produce heating of the metal by exciting the ions to vibrate about their fixed mean positions. Free electrons absorb higher-energy photons, hence the opacity of metals to visible radiation. If the incident photons are of even higher energy then some of the electrons may be emitted, but for solar energy conversion we can ignore this, since there is insufficient photon flux at such short wavelengths.

Quantum mechanics requires free electrons to be indistinguishable from one another, whilst the Pauli exclusion principle does not allow more than two electrons to occupy the same energy level (defined by four quantum numbers). As a result, there are a great number of possible energy levels in a metal, and to follow the changes in their occupancy it is necessary to use *statistical* mechanics. That is, we use the *probability* of a particular level being occupied. It is to be expected that the lowest energy levels are more likely to be occupied, unless the metal receives sufficient energy to raise electrons to empty higher levels. In fact, energy levels below a certain level, the *Fermi level* (associated with the *Fermi energy*), are completely occupied, and those above are empty at 0 K, with some smearing of this abrupt cut-off at higher temperatures. It is now possible to calculate the number of electrons which will be excited by a given energy input (and hence the absorption of the metal for any source), since only those electrons which have *empty* energy levels within reach will be able to gain energy. The energy which must be gained by an electron at the Fermi level for it to escape from the metal into a vacuum is called the 'work function'.

This simple free-electron model of metals successfully explains such features as their high electrical conductivity, but it is incorrect in detail and does not give the right specific heat capacities. If the theory is modified by the introduction of *allowed* energy 'zones' or 'bands' then better agreement is reached. The existence of allowed and forbidden energy levels arises through the periodic change in potential energy experienced by free electrons as they move through the array of ions.

3.2.2. *Semiconductors*

Some solids, whilst conducting electricity almost as well as metals, are *transparent* to some regions of the visible spectrum. In fact the absorption spectrum has detailed structure which may be attributed to energy absorption by fixed ions, free carriers and bound carriers in

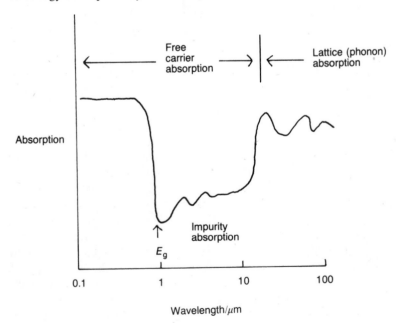

Figure 3.5. Optical absorption spectrum of a semiconductor.

a similar way to metallic absorption (figure 3.5). The most striking feature is perhaps the sharp change in absorptance at the energy corresponding to E_g, the bandgap energy, which for the semiconductors of interest in solar energy work lies in or near the visible waveband. The existence of a bandgap or forbidden energy gap is responsible for the major differences between metallic behaviour and semiconductor or insulator behaviour.

The energy bands within which free carriers are constrained arise in a similar way in *all* solids, but each has a different arrangement of bands. In metals the available energy bands overlap, and an electron at the bottom of a lower band can be excited by a photon to anywhere within the higher continuum of energies as long as there is room. In semiconductors and insulators the bands are separated, with the result that there is a range of energies which is forbidden to excited electrons. The difference between semiconductors and insulators is that insulators have a *wide* forbidden gap whereas semiconductors have a narrower gap which an electron can cross by absorbing a lower energy photon. This is summarized in figure 3.6.

The origin of these energy bands may also be understood by following what happens to a widely spaced assembly of atoms as they

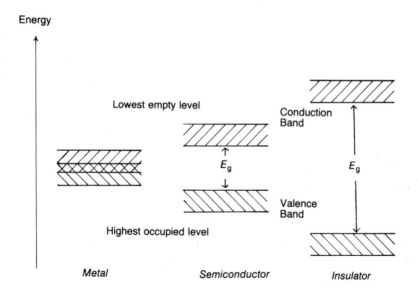

Figure 3.6. Energy bands in a solid at 0 K.

are brought closer together. This is known as the tight-binding approximation and leads to the same result as the free-electron approach. When an atom is isolated from all others, the electrons can take up only certain discrete amounts of energy, as shown schematically in figure 3.7 by the distinct horizontal lines. All atoms of the

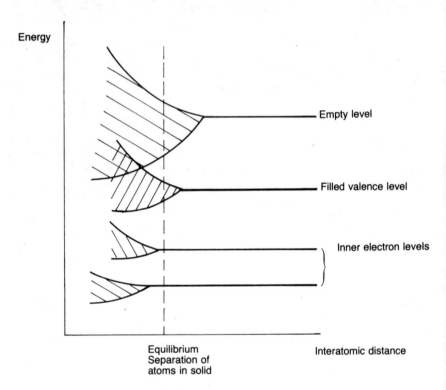

Figure 3.7. Electron energy levels in solids—tight binding model.

same sort will have these same energy levels as long as the interatomic distance is large, but as this distance is reduced these levels spread since the Pauli principle forbids two levels to be identical. A large collection of atoms forming a solid will have *bands* of very closely spaced energy levels corresponding to the total number of electrons and their total possible excited energy levels. The lower bands will be filled with electrons, and a quantum of energy can only be absorbed by the solid if it is sufficient to raise an electron to an empty higher level.

Figure 3.6 depicted the highest filled band and lowest empty band for our three classes of solids, at absolute zero, 0 K. This lower band is termed the *valence band* (VB) and the upper band is the *conduction band* (CB). At room temperature, a semiconductor will have a few free electrons in the conduction band, but most of the electrons will remain in the valence band, bound to the fixed ions. Rather than deal with a few free electrons in the CB and *many* electrons in the VB, it is customary in semiconductor physics to consider the *few* 'holes' left behind in the VB by the excited electrons rising to the CB, as effectively *positive charges*. Although these VB holes are a useful concept in understanding the behaviour of semiconductors, they do not quite behave as positively charged electrons, having a different effective mass from the free electrons in the CB. The absorption of photons of energy greater than E_g will excite more electrons into the CB, each leaving behind a hole in the VB. An externally applied electric field will produce a flow of electrons in one direction, and a flow of holes in the opposite direction. Of course, it is *possible* to consider the *electron* flow in CB *and* VB, but considerably simpler to deal only with electrons in CB and holes in VB. Electrical conduction in semiconductors is then due to the combined effects of moving charges in both bands.

3.2.3. *p–n junctions*

A very small concentration of foreign ions of the right kind incorporated into the regular array of host ions has extremely useful properties. The controlled addition of selected impurities is known as doping, and the foreign element is the dopant. The energy levels of an electron bound to one of the dopant ions will be different from those of electrons bound to the host semiconductor ions, with the most useful dopants having energy levels which fall within the bandgap. Each dopant atom added gives at least one discrete energy level within E_g. With a large concentration of dopant, these new levels would form a band of energies in the same way that CB and VB were produced, but this is not usually desired. If we can incorporate atoms from group V of the periodic table into silicon, these have an excess number of valence electrons with the consequence that they cannot all be tightly bound to the lattice. In other words, one electron in each dopant atom will be much more easily removed than the others. On our energy band model this means that they have energy levels very close to the CB. These extra donated electrons will make the semiconductor more electrically conducting than it was in the pure state, since we have effectively added some extra free electrons to it.

A similar result is obtained by adding atoms from group III. In this case there are too *few* electrons on each dopant nucleus, and another electron may be bound to the dopant to make up the deficiency. This is the same as saying that a positive hole has been *added* to the semiconductor, and such dopants accept electrons from the CB. The associated energy levels lie just above the VB.

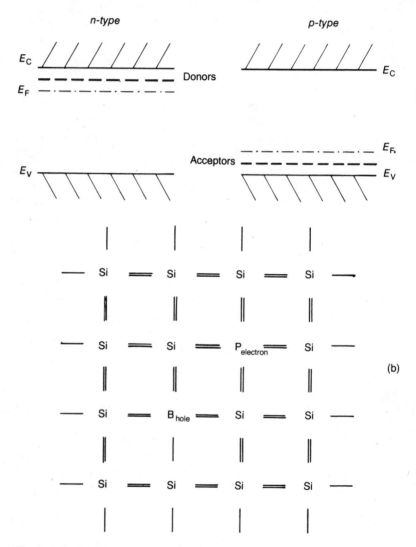

Figure 3.8. (*a*) Discrete energy levels produced by donors and acceptors in silicon. (*b*) Atomic bonding in doped silicon (two-dimensional model).

Figure 3.8 summarizes the position of these donor and acceptor energy levels within E_g, and schematically represents the bonding between silicon and a donor or acceptor atom. The conductivity type is called *n-type* when donors are added and *p-type* when acceptors are added. The material as a whole remains electrically neutral, since the additional free carriers are balanced electrically by fixed ionic charges.

The Fermi energy, E_F, lies within E_g for the devices we shall be considering, and it moves upwards or downwards according to the energy distribution of the electrons (figure 3.8). The *work function* of a semiconductor is defined as for a metal, but since there are few electrons at the Fermi level, it is more useful to deal with the energy required by an electron at the bottom of the conduction band for an escape into vacuum. This is called the electron affinity, $e\chi$.

For most semiconductor devices we are interested only in the absolute lowest energies in the CB and the absolute highest energies in the VB, although the energy bands may differ in various crystal directions. The only occasion when we shall refer to the CB minimum and VB maximum lying in different crystal directions is when we need to separate *direct gap* and *indirect gap* semiconductors. Direct gap semiconductors, like gallium arsenide, have the CB and VB closest together in the same crystal direction, whereas in an indirect gap material, like silicon, an electron may only be excited from VB to CB with the minimum energy E_g if a phonon also takes part in the process. Direct gap semiconductors, therefore, absorb radiation more strongly than do indirect gap semiconductors.

We can summarize the important features of semiconducting doping by reference to figure 3.9, which shows the temperature dependence of the free electron concentration. When the temperature is very low, there are few electrons in the CB, and the Fermi level is close to the CB. At higher temperatures there will be sufficient thermal energy to excite electrons from the donor levels into the CB, and a consequent falling of the Fermi level. A *small* change in temperature within this region AB has little effect on the conductivity, for each donor atom will have lost its loosely-bound electron, and the supply will be exhausted. This is the region within which stable semiconductor devices should be operated. A further increase in temperature will provide sufficient thermal energy to excite electrons across the gap from VB to CB, eventually in sufficient numbers to completely dominate the dopant electrons. The semiconductor then behaves as though it was undoped, with the Fermi level approximately half-way between VB and CB.

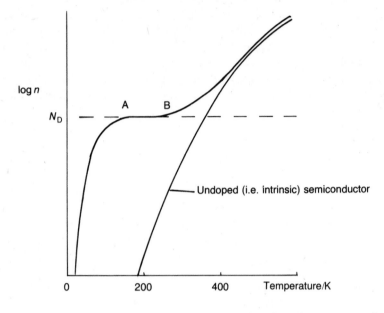

Figure 3.9. Free electron concentration, n, as a function of temperature for an n-type semiconductor.

If both donors and acceptors were added to the same piece of silicon, then one would compensate the other. A more interesting situation arises when *part* of the silicon is doped n-type and the adjacent part p-type, forming a p–n junction. This has associated with it a built-in electric field which arises as follows.

In *isolated* pieces of p- and n-type silicon there will be more free electrons in the n-region than in the p-region and vice versa for free holes. This uneven distribution in a p-n *junction* will tend to equalize by the diffusion of electrons and holes across the interface. Eventually this diffusion would be opposed by the electrical charges of the ionized donors and acceptors, and no further *net* diffusion would take place. As a result, a region on each side of the junction is depleted of free carriers and has a net positive or negative fixed charge. This charge sets up an electrostatic field which will separate any pairs of electrons and holes finding themselves within this 'depletion region' (figure 3.10).

The most helpful representation of this situation is the energy band diagram, showing the position of the Fermi level and conduction and valence bands, on each side of the junction. Any externally applied electric field will be added to the existing internal field at the

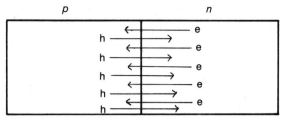

Diffusion of excess electrons and holes sets
up a depletion region.

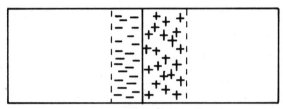

Fixed ionic charges, uncompensated locally by free
charges, give a built-in electric field.

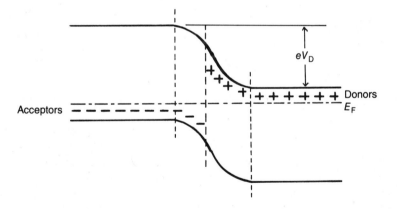

Figure 3.10. The p–n junction showing band-bending (zero bias).

junction, and the Fermi level will move accordingly. If it is not at the
same energy level across the p–n device, then a current will flow in
such a direction as to attempt to equalize the Fermi levels. This
means that with zero applied field the Fermi level must be constant,
and since it lies close to the CB in the n-type region and close to the

VB in the p-type region, the energy bands themselves must be bent at the junction (see figures 3.10). The band bending takes place within the depletion region, and indeed it is the flow of excess carriers initially producing the depletion region which also causes the bands to bend.

On this type of diagram, the tendency for free carriers to seek the lowest energy possible is shown by electrons 'sinking' downwards and holes 'floating' upwards. Consider any conduction band electrons in the n-region, where these are in the majority: if they are to enter the p-region CB they must gain sufficient energy to overcome the barrier, eV_D. (Holes in the p-region have an exactly similar barrier in the valence band to overcome.) On the other hand, any electrons in the conduction band of the *p-region*, where they are in the minority, can pass unimpeded across the junction, for the energy step is in their favour. We can see that any free electrons and holes produced by the absorption of photons will be separated by the field due to this potential barrier, once they enter the depletion region: electrons moving 'downhill' to the n-side and holes moving 'uphill' to the p-side. The limiting part of this process is ensuring that the *minority* carriers have time to diffuse into the depletion region before they are attracted by the oppositely charged *majority* carriers with which they are surrounded, and with which they will recombine (i.e. a loss of free carriers).

We have avoided introducing any equations or quantitative treatment of the p–n junction, although this has meant that a slightly wordy derivation has been necessary. Before concluding this chapter and going on to discuss ways of applying these theories of the interaction of light with matter, it is necessary to add a single equation for future reference. This is the formula for the diffusion current which flows in a p–n junction, and which is responsible for the initial bending of the CB and VB. It is found that this current flows even when the addition of a large external negative voltage to the junction causes the potential barrier to be so high as to block all other current. In other words, the p–n junction will have a small leakage current in reverse bias, when an ideal device would be an effective open circuit. This reverse bias saturation current, or leakage current is I_{01}.

$$I_{01} = en_i^2 \left(\frac{1}{N_D} \frac{L_p}{\tau_p} + \frac{1}{N_A} \frac{L_n}{\tau_n} \right) \tag{3.9}$$

n_i is the concentration of free carriers in the pure semiconductor, and N_D, N_A are the net dopant concentrations in n and p regions, respectively. L is the minority carrier diffusion length for each side of the junction, and is a measure of the distance travelled before recombination occurs. τ is the corresponding minority carrier lifetime. From this equation we can see that a junction with minimum leakage current should be highly doped, with a low intrinsic carrier concentration n_i, other quantities being equal. It also indicates that if n_i should increase, for instance by overheating, the p–n junction will conduct heavily.

4. The photothermal effect

The most obvious exploitation of the Sun's radiation is for heating. In all such applications the efficiency of the process depends on the quality of the solar absorber. Photothermal conversion uses the absorption not only of infra-red but also of visible radiation. Indeed, good photothermal collectors may *reject* much of the solar infra-red spectrum and be effective with only the wavelengths shorter than about 2 μm, though only about 5% of the solar radiation is rejected in such a case.

The spectral distribution of solar radiation suggests that an effective collector would be one which absorbs strongly throughout the visible region of the solar spectrum, whilst losing the minimum of energy by conduction and convection to the surroundings, and which radiates as little as possible in the infra-red region of the spectrum. As the temperature of this collector rises, its rate of loss of energy by radiation increases according to the Stefan–Boltzmann law. Figure 4.1 shows a body receiving solar energy, and the many energy paths associated with it. Above about 100° C the energy losses by conduction and convection become less important than that by radiation. An equilibrium temperature is reached when there is a balance between solar input and heat loss. Below 700° C the radiant loss occurs mostly at wavelengths greater than 2 μm (figure 4.2).

4.1. *Selective absorbers*

The idea of a *spectrally selective absorber* now arises. If we operate a collector at a temperature of, say, 700° C then the loss will be mainly by radiation in a waveband different from that of the incident solar radiation, the boundary between the two bands being at about 2 μm. Since at any wavelength absorptance is equal to emittance, for a black body at constant temperature, it is impossible to decrease its emittance without decreasing its absorptance. However, the absorptance need only be high for the band from 0·2 to 2 μm, and emittance should be low only for wavelengths greater than 2 μm. A suitable

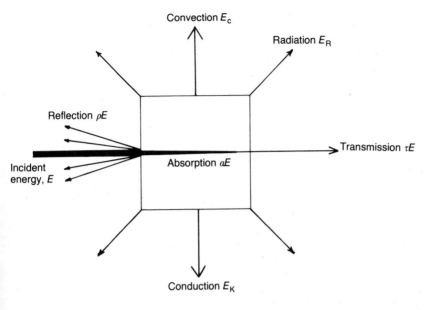

Figure 4.1. Energy balance of a body receiving solar energy, E.

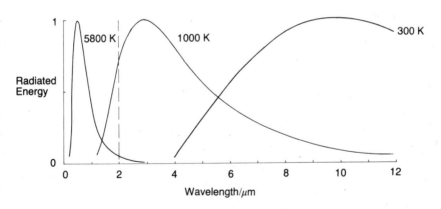

Figure 4.2. Radiant energy loss of a black body at selected temperatures.

selective absorber will have the characteristic shown in figure 4.3, and
will behave as shown in figure 4.4, reaching a higher equilibrium
temperature than a black body as long as it retains high absorptance.
This may be demonstrated by equating the radiant loss (σT^4) per unit

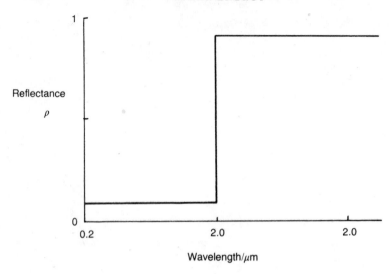

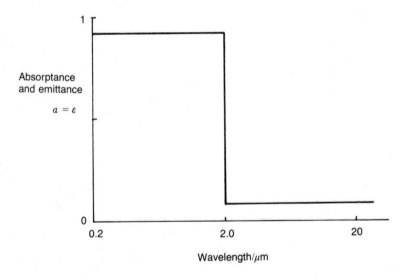

Figure 4.3. Characteristics of a selective absorber (a 'dark mirror').

area to the absorptance of solar energy (aE) for both a black body, and a selective absorber with a(visible) $= 1\cdot0$ and ε(infra-red) $= 0\cdot1$.
    At equilibrium:

$$aE = \varepsilon\sigma T_E^4 \qquad (4.1)$$

For a black body
$$T_E^4 = \frac{a}{\varepsilon} \, \frac{E}{\sigma} = \frac{E}{\sigma} \qquad (4.2)$$

For a selective absorber:
$$T_E^4 = \frac{a(\text{visible})}{\varepsilon(\text{infra-red})} \, \frac{E}{\sigma} \qquad (4.3)$$

$$= 10\frac{E}{\sigma}$$

If E is 1 kW m^{-2} then the equilibrium temperatures are 364 and 648 K, respectively.

Some words of caution are necessary regarding the ratio $a(\text{visible})/\varepsilon(\text{infra-red})$. Although the above expressions suggest that a value of $a(\text{visible})$ of 0·5, say, together with a value of $\varepsilon(\text{infra-red})$ of 0·05 (ratio 10 : 1) should lead to the same equilibrium temperature (648 K), we have ignored convective and conductive heat losses, and the extraction of heat from the collector. The inclusion of these in the calculation would show that it is important that, as well as the ratio $a(\text{visible})$ to $\varepsilon(\text{infra-red})$, the value of $a(\text{visible})$ itself is as high as possible. People often omit the suffix 'visible' or 'infra-red', for it is common practice to use a and ε only in this context.

Real selective surfaces can approach the ideal behaviour of figure 4.3, but the steeper the edge separating high- and low-absorptance regions, the less perfect is the wavelength-independence of $a(\text{visible})$ and $\varepsilon(\text{infra-red})$. This fine-structure in the spectral reflectance of real surfaces may be removed by additional coatings on the surface, but some overlap of the two regions is unavoidable. Even with the most abrupt step which has been produced, some solar infra-red radiation is rejected, and some shorter wavelength radiation is re-radiated with high emittance. Values for $a(\text{visible})/\varepsilon(\text{infra-red})$ of 12–15 are attainable, even at the elevated operating temperatures (see table 4.1).

The absorptance/emittance ratio of black paint compared with green or white paint would suggest at first thought that it is better to paint space-heating radiators black. Unfortunately, there is little improvement to the heating efficiency with this treatment because heat transfer takes place mainly by convection rather than by radiation at the usual operating temperature! In any case the *infra-red* emittance of these paints is the same! It is important to include *all* energy transfer processes when estimating the benefit to be gained from selective coatings.

None of the *simple* surfaces listed in the table give values of absorptance/emittance near 12–15. It is the more recent (and com-

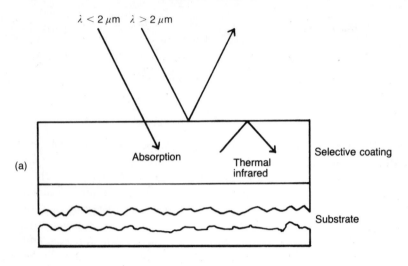

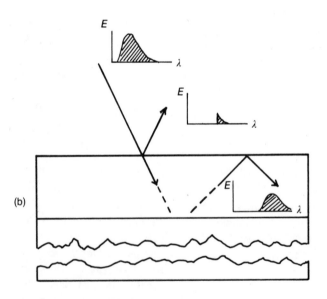

Figure 4.4 Behaviour of a selective coating on an absorber exposed to sunlight. (a) Light of wavelength <2 μm is absorbed, of wavelength >2 μm is reflected, so that (b) solar radiation is divided, with only the longer wavelengths reflected.

TABLE 4.1. *Absorptance and emittance values for various materials.*

Surface	Temperature $\theta/°C$	a (visible)	ε (infra-red)
Ice (snow)	0	0·31–0·33	0·96–0·82
Black matt paint	20	0·95	0·95
Green paint	20	0·50	0·95
White paint	20	0·25	0·95
Lampblack	20	0·95	0·89
Polished copper	200	0·35	0·04
Aluminium	20	0·09	0·10
Silver	200	0·04	0·04
Copper (oxidized)	20	0·90	0·14
Nickel black Chrome black (i.e. (Ni + Zn + S): Depends on substrate)	20	0·89–0·95	0·16–0·07

plex) surface treatments which yield such high values. The behaviour of metal surfaces, which already exhibit low infra-red emittance, is improved by an overlayer of a semiconductor which is highly absorbing in the visible, but which is thin enough to be transparent to infra-red wavelengths and so retain the low emittance of the underlying metal. Suitable semiconductors with high a(visible) tend to have high refractive indices and are, therefore, a bad match to the low refractive index of air. Unless an anti-reflection coating is added to the top surface of the semiconductor–metal 'tandem' the reflectance is undesirably high. One way to avoid this is to use semiconductors prepared so as to contain a large concentration of tiny voids. These reduce the refractive index (from 4·1 to less than 2 for lead sulphide) without making them transparent to visible radiation.

More complex absorber coatings have been constructed to improve selectivity and to tune the absorption edge to exactly the optimum wavelength for the collector temperature in mind. These use the phenomenon of interference within an absorber film bounded by two thin metal films. The visible light which passes through the top (semi-transparent) metal layer makes many passes across the absorbing film due to internal reflections (see figure 4.5). A second dielectric film on the top surface of the stack broadens the wavelength region of high absorption. Variations on this structure

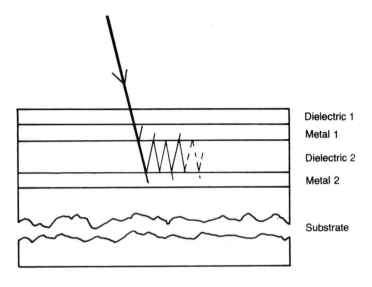

Figure 4.5. Selective absorber using interference.

have most of the absorption within the lower metal layer and the dielectric above does not absorb strongly. These stacks retain the high infra-red reflectance of the metal substrate. The exact structure and composition of each thin layer is almost certainly different from the succession of materials as they are laid down, since interfacial diffusion takes place, enhanced by an increase in temperature. All interference-type absorbers are liable to degradation by this process and therein lies their weakness for long-term operation at elevated temperatures. It may be difficult to manufacture large-area selective absorbers reproducibly and reasonably cheaply. Note that the desired interference effects only operate for near-normal incidence of the solar radiation.

An improvement on conventional vacuum deposition of the multi-layer is to use chemical vapour deposition. By this process, silicon compounds, say, can be sequentially deposited on a substrate by changing the composition of the gaseous compound fed into the heated reaction chamber. A wide acceptance angle anti-reflection coating could be made by depositing silicon–oxygen–nitrogen compounds on to the metal, arranging them to have a gradually changing refractive index through the layer.

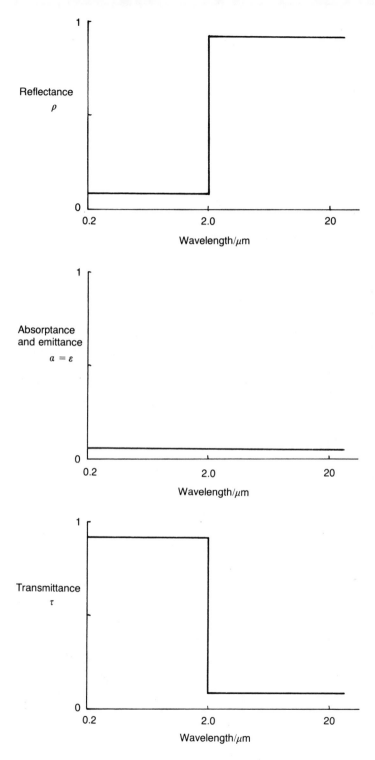

Figure 4.6. Characteristics of a selective transmitter ('heat mirror').

4.2. *Selective transmitters*

Transparent selective coatings deposited on the inside or outside surface of the cover plate of a solar collector, transmit visible radiation from the Sun, but reflect the infra-red radiation from the heated absorber plate back towards the absorber (see figure 4.6). They are sometimes called heat mirrors. (In contrast there exist cold mirrors which reflect visible and not infra-red radiation. These would be preferred to normal metallized reflectors for photovoltaic cells since the cells will only receive the energy band which they can absorb.). Even glass itself is a selective transmitter since it will absorb infra-red more than visible radiation, and special glasses are used as heat filters in some optical equipment. The noble metals, silver and gold, can be applied as thin surface films (20 nm) to glass to improve its selective transmittance. These absorb some of the visible region of the spectrum but are slightly better than metals such as copper. Even better coatings are based on indium–tin oxides. Figure 4.7 shows how a selective transmittance coating can be used.

A glass enclosure can form a heat trap by the 'greenhouse' effect, for the structure will allow visible solar radiation to enter and be absorbed within, but will not allow the radiant energy from the

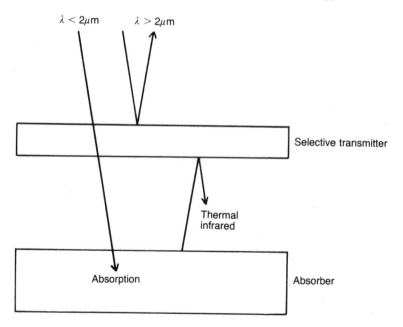

Figure 4.7. Behaviour of a selective transmitter above a solar absorber.

heated interior to pass out. There is a reduction in the energy loss with the same result as using a selective absorber: a high equilibrium temperature is reached. The same effect is produced even when the glass is replaced by sheets of an infra-red transparent plastic, or rocksalt, which suggests that the behaviour of the enclosure has not been correctly modelled, for an important feature of the enclosure is to reduce conductive and convective heat transfer to the atmosphere from the absorber. We are dealing with a situation in which radiative heat loss is only one of several paths.

The selective transmittance of the Earth's atmosphere is of importance for solar energy applications, and much more so for maintaining a reasonably high night temperature on the surface. If the atmosphere was transparent to infra-red radiation from the heated Earth's surface then the night-time temperature would always fall by radiative cooling. Certainly when there is no cloud cover, the surface temperature *can* fall to below 0° C, but the presence of water vapour and other gases reduces this net cooling.

Some confusion exists in the literature of selective surfaces as to the differences between selective *absorber* coatings and selective *transmittance* coatings, since either can be applied directly to the absorber surface. Both have *high* infra-red reflectance, but selective absorbers block the visible radiation whilst selective transmitters pass on the visible radiation to an underlying absorber. (Compare figures 4.3 and 4.4 with figures 4.6 and 4.7.) Selective transmittance coatings applied to a cover glass keep much cooler than selective absorber coatings which are in contact with the high-temperature absorber itself.

One example of a multilayer selective transparent coating applied to the inside of the cover glass of a flat-plate solar collector is a sandwich of 13 nm silver between two 33 nm layers of titanium dioxide. It has 80% transmission in the visible and 95% reflectivity in the infra-red at 1000 nm. If used in conjunction with a selective absorber having $a(\text{visible}) = 0.1$ and $\varepsilon(\text{infra-red}) = 0.2$, then the calculated thermodynamic efficiency for the electrical conversion of solar energy is 25%, using a Carnot cycle heat engine operating at 750 K. Later we shall see whether or not these conditions are reasonable.

The efficiency of these coatings may be improved by reducing any reflection losses at the top surface with an additional layer of dielectric of selected refractive index. Since anti-reflection coatings of a single thin film are effective only for a narrow band of wavelengths, and for near-normal incidence, a collector using them may have to track the Sun. Multilayer surface coatings are more effective, but the additional complexity may not justify the energy saving.

4.3. *Size effects (wavefront discrimination)*

A novel approach to improving the performance of selective transmittance coatings has been suggested recently. Openings of the same dimensions as the cut-off wavelength are etched through the film to make a micro-mesh. The openings will transmit shorter wavelengths than their aperture but will not be 'seen' by longer wavelengths: to infra-red radiation the mesh appears smooth and continuous. This is closely analogous to the propagation of microwave radiation through waveguides, and the theory has been developed to the point at which infra-red mesh filters can be designed and constructed. Large-area selective meshes for solar energy converters have yet to be successfully demonstrated.

The absorption of visible radiation may also be enhanced by roughening the surface of an absorber, which will generally increase the absorbing area and cause multiple reflections among the surface features. If the roughness is on a small enough scale the surface will still appear smooth and reflecting to longer wavelengths, thus an increase in the ratio a(visible)/ε(infra-red) is possible. Even a dense array of needle-like projections from the surface (e.g. metal whiskers) will improve the absorption of solar radiation by scattering light downwards rather than upwards away from the collector. The infra-red emittance of these surfaces is generally increased as well, and their long-term performance may suffer from mechanical damage.

The older metal 'blacks', such as gold or platinum black, are more complex in their selective absorption mechanism than merely providing a roughened, low reflectance surface. These preparations of small metal particles on a reflecting substrate, or in a host matrix, resonantly scatter radiation of wavelength comparable to the particle size. Depending on their preparation, there is either selective absorption or selective transmission of the infra-red, together with strong absorption in the visible. Protection of special textured surfaces from contact or weathering is essential since they can be damaged even by a light touch.

5. Solar concentrators

We can use selective surfaces to reduce radiative losses, and hence achieve high temperatures, but for the highest temperatures we *must* increase the input energy density by concentrating the solar radiation. In a later chapter we shall compare the influence of a highly concentrating optical system on the equilibrium temperature of an isolated collector, with the temperatures achieved by simple flat-plate

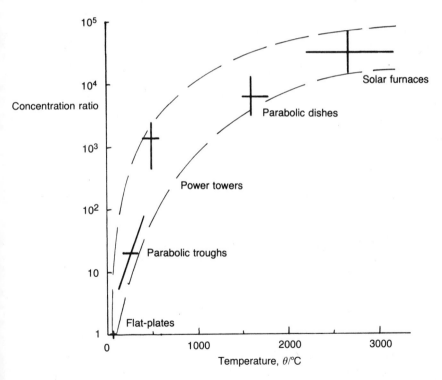

Figure 5.1. Temperatures reached by solar absorbers with concentrator optics.

collectors. Figure 5.1 summarizes the temperatures achieved with solar concentrator optics.

5.1. *Some possibilities*

The difficulties of making large refracting optical elements suggest that mirrors, rather than lenses, will be important for photothermal collectors, although experiments with plastic water-filled lenses suggest a possible alternative. 'Fresnel lenses' (figure 5.2) have been

Figure 5.2. Cross-section through a Fresnel lens. (Aberrations are lowest if illumination is incident on the profiled surface.)

used for cheap concentrators for small photovoltaic cells. It is possible to make them from extruded plastic in a linear form for larger areas. The accumulation of dirt reduces their performance.

We define the concentration ratio (CR) of a mirror system as the ratio of aperture area to absorber area, assuming that the absorber intercepts all reflected rays. Thus four plane mirrors each inclined outwards at 120° to a flat-plate collector of the same area as each mirror will give a CR of 3 for the collector as a whole. The effect of a high CR is reduced from the simple calculated value by the actual reflectance of the mirror surfaces, which may be low for shallow angles of incidence. Diffuse (scattered) solar radiation arrives on a surface from all angles and so cannot be focused—only the direct radiation from the Sun can be focused to an image. These collectors would be a poor choice for UK latitudes.

Concentrator ratios of more than 1000 are possible with parabolic dishes, but these must track the Sun accurately for full effectiveness. A tracking mechanism (probably using a dual combination of timing and photosensing) able to withstand wind and weather may double the cost of the basic absorber. Maintenance of the high optical perfection of the surfaces is added to the operating costs.

Even simply curved mirrors with low values of CR can be used in a large array to produce high temperatures at a common focus. The number and quality of mirrors needed to supply sufficient solar energy to melt tungsten is accurately known, since a 1 MW (thermal power) solar furnace of this magnitude has been operated in the

French Pyrenees at Odeillo. This follows an earlier 50 kW French construction at Mont-Louis. The newer design is an array of 63 plane mirrors, each of 45 m² and containing 180 mirror panes, which track the Sun and reflect its image southwards to a 54 m × 40 m paraboloidal reflector. This fixed vertical paraboloid is made of 9500 flat-plate mirrors bent by mechanical constraints to give a better image than flat mirrors. The focal plane is 18 m from the paraboloid and temperatures of up to 4000° C are reached in a 50-cm² hot spot. High temperature metallurgy remains the main use of this installation, although the cavity collectors which are being designed for the American 'power tower' scheme are being tested there.

Somewhere between these extremes there is a CR sufficient to achieve a 500° C working fluid temperature for efficient solar electricity generation. The losses at these levels of concentration will be both radiative and conductive/convective. Many experimental arrangements have been built, and some successfully operated for years, since the mid 1800s. Generally these were used to operate steam engines, for instance to irrigate fields in the Tropics. Between 1860 and 1880, Professor A. B. Mouchot built several solar-driven irrigation pumps and demonstrated the effectiveness of the method in Algiers, raising 2½ tons of water per minute by what is now known as an axicon concentrator. This was a truncated reflecting cone of mirrors focusing sunlight on to a tubular boiler along the axis. The dish was tilted towards the Sun and was moved around to follow the Sun. Despite poor optical surfaces and high heat losses in the fluid cycle then used, the results were promising enough to trigger a spate of patent applications and the founding of solar power companies. One of these was Boys' and Shuman's 'Eastern Sun Power Corporation' in the USA which built a number of high-temperature collectors based on flat- and curved-mirror troughs. The most celebrated of their solar engines was the one built at Meadi in Egypt which operated successfully for years at an overall efficiency of about 5%, supplying irrigation water. The collectors were troughs 62·5 m long, with a projected area of 1260 m², and rotated about their long axis to follow the Sun (figure 5.3). Then, as now, the cost of solar power plants was higher than conventional fossil fuel plants, and reductions in cost and increases in efficiency are still awaited!

5.2. *Parabolic collectors*

In the most common types of parabolic trough collector, the image size is approximately given by the product of reflector focal length and angular diameter of the Sun, which gives some idea of the

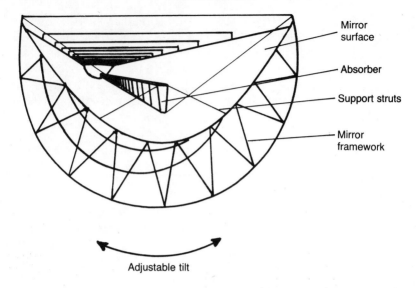

Mirror
surface

Absorber

Support struts

Mirror
framework

Adjustable tilt

Figure 5.3. A view along one of Shuman's trough collectors, used in Egypt, 1913.

absorber size needed, if not its shape. A large cross-section absorber will obscure the Sun from a considerable area of the reflector, and will probably need strong rigid supports along its length to avoid distortion. The conductive heat loss along these is far from negligible.

The shape of the absorber is important, for areas of a parabolic reflector near its rim will reflect light obliquely on a plane through the focus, so the image will not be uniformly intense unless the absorber profile matches its shape. Although the absorber is often in the form of a pipe containing the fluid to be heated, another pattern is the cavity. At high working temperatures, a well-designed solar collector will lose heat mainly by radiation. If the absorber is made in the form of a re-entrant body with interior baffles, the reflected solar energy passes into a cavity through a small aperture and is absorbed within (figure 5.4). The size of the aperture is related to the losses, but must be made larger than simple calculations would suggest, to allow for imperfections in the reflector surface. Nonetheless, a cavity absorber can only be used with small rim angle reflectors.

There is no practical advantage in increasing the rim angle, the angle subtended at the focus by the edges of the mirror, for the useful energy from the edges of even a perfect profile reflector shows a diminishing return for each increment of angle. Each small increase made to the rim angle is an extra contribution to the total area of the

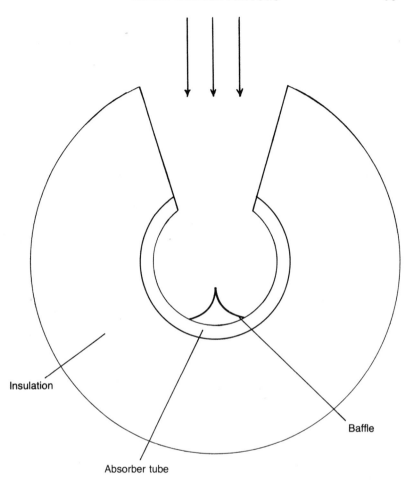

Figure 5.4. Cross-section through a cavity absorber.

mirror (figure 5.5(b)) but the reflected *oblique* image on a plane absorber from this additional area will counteract this gain. Theory suggests a CR for parabolic dishes of $46 \cdot 1 \times 10^3 \sin^2 \theta$ for a rim angle θ, but the unavoidable imperfections in large parabolic dishes are usually worse towards the edges and this sets the upper limit on the performance. For real parabolic dishes the CR is limited to about 10^3, which still allows absorber stagnation temperatures to approach 5000 K. The inherent low heat capacity of these concentrator collectors with high CRs, together with their dependence on tracking mechanisms, makes them undesirable for many applications.

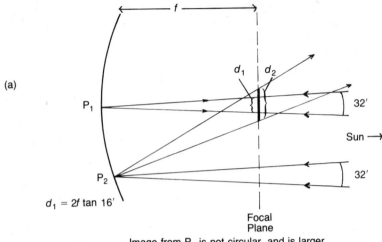

$d_1 = 2f \tan 16'$

Focal
Plane

Image from P_2 is not circular, and is larger
than that from P_1

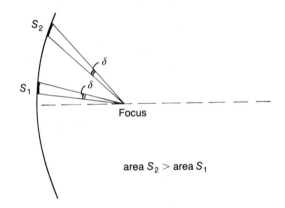

area $S_2 >$ area S_1

Figure 5.5. Parabolic mirror geometry and sizing.

5.3. *Winston collectors*

A new class of concentrator which has ideal optical behaviour (but
which is not necessarily ideal for solar energy collection) has been
developed by R. Winston. Originally intended for use with Čerenkov
radiation detectors for high-energy particle physics, the designs have
been scaled up to the sizes needed for solar purposes. These
concentrators are called 'ideal' because all radiation entering within a
certain acceptance angle, θ_A, will be directed through an aperture at
the bottom of this optical funnel, and no radiation outside this angle

will reach the aperture. They are not imaging collectors for there is no focal point. One two-dimensional shape in this class is derived from two separate parabolas, hence its name of Compound Parabolic Concentrator (CPC) (figure 5.6). The three-dimensional surface formed by two CPC collectors joined at right angles, to make a cup shape, is *not* ideal in the optical sense given above, but the surface formed by rotation about the axis is.

A further property of an ideal collector is that radiation emitted from a correctly placed absorber will all be returned, through the acceptance angle, to the source. This means that the radiative loss out of the CPC is the least possible for any concentrator. The convective loss may be very high unless the absorber geometry is arranged to prevent warm air currents from rising out of the funnel mouth (figure 5.7). A cover plate over the mouth of the funnel will reduce convection currents and will protect the very large area of optical surface. The large area of CPCs is a disadvantage, but they do have a higher CR than any other collector with the same acceptance angle which means that the Sun can change its angle in the sky much more for CPCs than for other concentrator collectors before reorientation is necessary.

The CR of an ideal concentrator is

$$CR_{ideal} = n_2/n_1 \sin \theta_A \quad \text{(for a two-dimensional collector)} \quad (5.1)$$

(The right hand side of the equation is squared for a three-dimensional concentrator.) This is equivalent to the Carnot efficiency for heat engines. The refractive indices, n_1 and n_2, refer to the media outside and inside the collector, and will often be equal. For the Winston CPC the CR is given by the ratio d_1/d_2 (figure 5.6). The length of the CPC is

$$L = \tfrac{1}{2}(d_1 + d_2) \cot \theta_A \quad (5.2)$$

This shows that high CRs require a very deep collector. An advantage of this collector over a simple V-shaped trough is that multiple reflections are fewer, which gives some tolerance in the surface reflectance. Since reflectivity losses can be 5–10%, the energy absorbed by the reflectors could raise their temperature appreciably. Cooling the mirrors themselves would give a second source of thermal energy at a lower temperature than that from the proper absorber at the exit aperture. It may not be a disadvantage to allow the reflectors to heat up, since the absorber is losing heat by radiation and the net radiant loss depends on the temperature difference between source and sink.

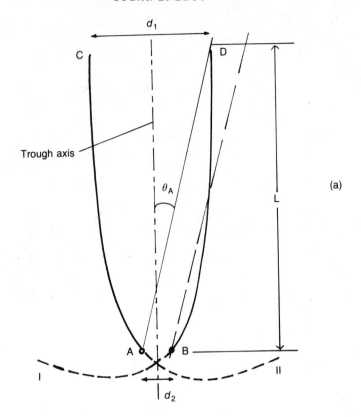

(a)

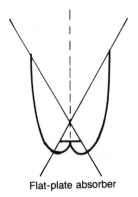

Flat-plate absorber

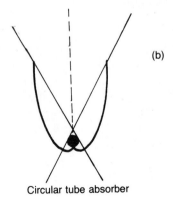

(b)

Circular tube absorber

(c)

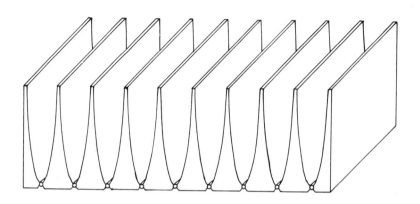

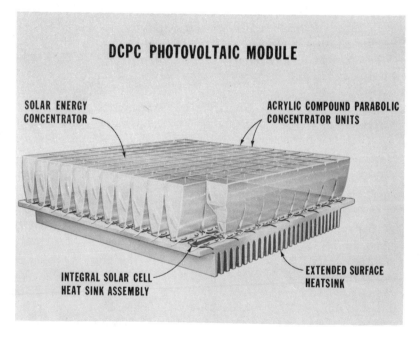

Figure 5.6. The Winston compound parabolic concentrator (CPC). (*a*) A is the focus of the parabola I, B is the focus of parabola II. Mirror surfaces are parallel to trough axis at C and D. (*b*) Terminations to the CPC to match absorbers. (*c*) Large array of CPCs with tube absorbers. (*d*) A schematic diagram of cast CPCs with a solar-cell array. (Reproduced by kind permission of the Argonne National Laboratory, USA.)

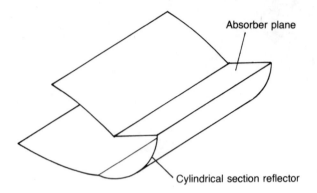

Figure 5.7. Low convection-loss absorber for a CPC.

If the collector is not to follow the Sun, then it must have a large acceptance angle (i.e. small CR). Most concentrator collectors will then collect a small amount of diffuse, scattered sunlight but will perform badly in overcast conditions compared with a flat-plate collector. The CPC will collect a fraction $1/CR = \sin \theta_A$ of this scattered light and so does not do as badly as highly concentrating parabolic arrangements. The collection of diffuse light can be improved by truncating the CPC at the entrance aperture, which increases the effective acceptance angle (although it also decreases the CR). It will then operate for a longer period each day. Thus,

$\theta_A = 6°$ gives $CR = 9\cdot6$ with 8 h/day collection;

$\theta_A = 2°$ gives $CR = 28\cdot6$ with 4·4 h/day collection.

With the CPC a flat-plate parallel to the lower aperture will collect all the exit energy, but better absorber shapes can be used to match the collector as well as to reduce conduction/convection losses. Radiative losses can only be reduced by selective absorber coatings. An asymmetric ideal concentrator of sea-shell form has been suggested using a cavity collector at the exit to reduce convection (figure 5.8).

The CPC should match the particular cross-sectional shape of the absorber tube, if the radiation is to heat fluid in a pipe. This means that the end of the CPC should form a suitable curve surrounding the absorber tube (see figure 5.6). The convex absorber can be of any shape (circular, oval or 'flat' cross-section) provided that the circumference is $d_1 \sin \theta_A$, and that it does not require radiation to be reflected at shallow angles of incidence on the CPC.

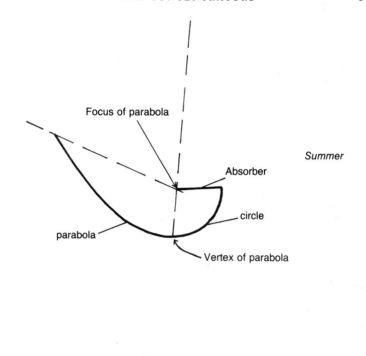

Focus of parabola

Summer

Absorber

circle

parabola

Vertex of parabola

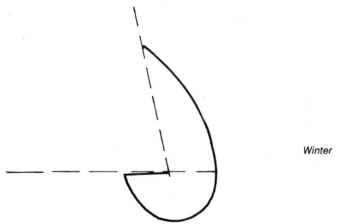

Winter

Figure 5.8. Cross-section through A. Rabl's non-tracking 'sea-shell' concentrator collector.

Let us examine the theoretical maximum CR of the CPC for use as a solar collector. The Sun being a disc with angular half-width of 32′, the minimum acceptance angle for collection of all direct insolation is

16' if the mirror geometry is perfect. Hence a two-dimensional CPC should, in theory, be capable of a CR of up to 215. This implies a maximum possible absorber temperature of about 1400 K, if only radiative transfer of heat from the absorber is allowed.

Higher concentrations than this are achieved by parabolic dish reflectors, or by large arrays of plane mirrors, but the absorber used with these must be carefully constructed to make full use of the high energy flux. It may sometimes be advantageous to use a tandem concentrator—say an array of plane mirrors, with a CPC collecting the reflected radiation for a cavity absorber. The cost of producing specially profiled mirrors in quantity must influence the decision as to the best choice of collector for any given application. If the elimination of surface errors in curved mirrors is expensive it may be better to use high quality *plane* mirrors to 'synthesize' the shape required.

6. Photochemical reactions

6.1. *Introduction*

The maximum thermodynamic efficiency of an engine operating with the Sun as source and a 300 K sink is as high as 95%, but the only way to approach this efficiency would be to use the high quantum energy of the photons from the Sun to produce electrical or chemical energy rather than to convert all photons to low-grade heat. Photosynthesis and photovoltaic converters do this, although their limiting efficiencies fall far short of 95% due to a photon threshold below which no radiation is absorbed. Similarly, a photochemical reaction can produce usable chemical fuel, or can be used in a photogalvanic (or photoelectrochemical) cell to produce electrical energy. It is necessary to distinguish between photochemical energy conversion and photocatalytic processes. A photocatalytic reaction is thermodynamically possible, but its rate is restricted by a kinetic barrier which can be reduced by photon absorption. Photochemical energy conversion occurs only when illumination changes the thermodynamic balance between products and reactants (by reducing the energy level of one of the reacting components) and so allows the reaction to proceed (figure 6.1).

6.2. *Photochemical conversion*

We require a non-corrosive, stable photochemical with high solar absorption (preferably black) which would produce a suitable stable chemical product with a high quantum yield. Our ideal reaction would then require no *thermal* storage of energy, so eliminating many of the losses inherent in photothermal collection. Unfortunately, most photochemical reactions proceed only under rather intense ultraviolet illumination. Several reactions which might be considered are very inefficient even in monochromatic light of the optimum wavelength, since they produce unstable products, which either revert rapidly to the initial reactants, or take part in side-reactions giving useless products. If a reaction with a low threshold energy

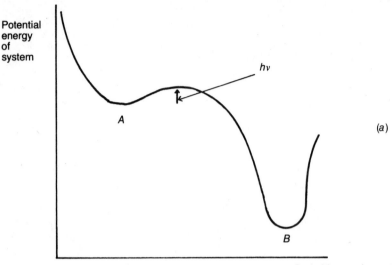

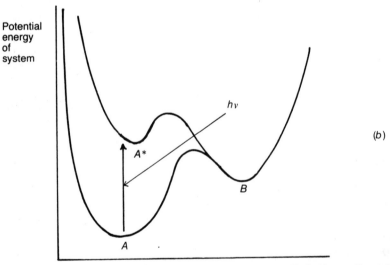

Figure 6.1. The difference between photocatalysis and a photochemical reaction. (a) Photocatalysis: $A \xrightarrow{h\nu} B$; (b) photochemical reaction: $A \xrightarrow{h\nu} A^* \to B$.

could be found, then even under overcast conditions there would still be useful energy conversion. On absorption of a photon of sufficient energy, a molecule of photochemical will be raised to an excited state, followed rapidly by transfer of this energy to dissociate the compound, or to initiate a reaction, or a less desirable process such as re-emission of photons (fluorescence). An incident photon with energy below the visible will only cause vibrational or rotational excitation.

An ideal component of a photochemical collector of solar radiation would be water, if it could be made to fulfil certain requirements, for it is readily and cheaply available, and can be converted to hydrogen given sufficient energy input. The *thermal* dissociation of water needs temperatures of 2500° C, since the threshold energy is high. This suggests that photodissociation would be more easily handled. There has been extensive research into the 'sensitized' photolysis of water in an attempt to reduce the threshold energy for hydrogen production to less than 1 eV: the complete photodissociation of liquid water into molecular oxygen and molecular hydrogen needs photons of energy greater than 2·5 eV. Water without an additive is not a good solar collector of even high-energy photons, as evidenced by its transparency!

$$H_2O \xrightarrow{h\nu} H_2 + \tfrac{1}{2}O_2 \qquad (6.1)$$
$$\text{(liquid)} \qquad \text{(gas)} \qquad \text{(gas)}$$

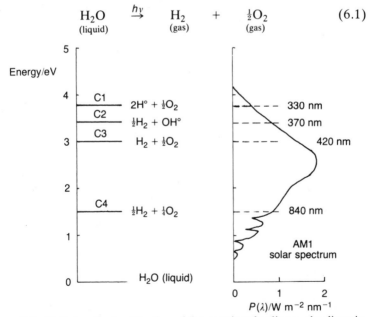

Figure 6.2. V. Balzani's classification of the reactions leading to the dissociation of water, compared with the spectral energy distribution of solar radiation (see figure 2.1, p.11).

The above equation represents the 'C3-type' photodissociation of liquid water, which is a path requiring 3 eV photons. This and other paths are represented on an energy scale in figure 6.2. The production of a single molecule of oxygen together with two molecules of hydrogen requires two molecules of water, and the transfer of four electrons. Photodissociation is a multi-particle process and so has a low inherent probability. Obviously path C4 is to be preferred on energy grounds alone, since 40% of the AMI solar spectrum has sufficient energy to drive this reaction. We must now see which, if any, of these possible routes might be used to produce hydrogen from water with solar radiation alone.

Attempts to discover the low threshold energy reaction which would allow high efficiency solar photolysis of water go back to the 1950s. The reaction rate of the C3 process is increased by adding cerium perchlorates, although these do not reduce the threshold energy. In the correct proportions these compounds give a balanced series of reactions with the net production of molecular hydrogen and oxygen. The possible steps in this process are given below:

$$\text{(oxidation)} \quad 4Ce^{4+} + 2H_2O \rightarrow 4Ce^{3+} + 4H^+ + O_2 \qquad (6.2)$$

$$\text{(reduction)} \quad 4Ce^{3+} + 4H_2O \rightarrow 4Ce^{4+} + 4OH^- + 2H_2 \qquad (6.3)$$

The net effect of this reduction/oxidation pair is given by:

$$6H_2O \rightarrow 4H^+ + 4OH^- + 2H_2 + O_2 \qquad (6.4)$$

The free radicals, represented in this simplified scheme by H^+ and OH^-, most readily combine to form water, with the overall result that reaction (6.2), the C1 reaction, and reaction (6.3), the C2 reaction, together represent a C3 reaction:

$$2H_2O \rightarrow 2H_2 + O_2 \qquad (6.5)$$

In order for this net C3 effect to occur, reactions (6.2) and (6.3) must be balanced, by using the correct proportion of the reducing cation (Ce^{3+}) to the oxidizing cation (Ce^{4+}). Reaction (6.3) has a lower energy threshold than reaction (6.2), as can be seen from figure 6.2, and as a result it has a lower quantum yield.

These reactions are easily poisoned by impurities, but a more serious objection is that both proceed via formation of free radicals, which are highly reactive. Side-reactions with these can drain away the stored photochemical energy. A further energy leak is caused by the back-reaction of the products if they are not separated.

In more general terms, the type of process needed for solar energy conversion is termed 'endergonic', i.e. one in which the reactants will be converted when illuminated to products with higher bond energy. The products should be stable until activated (e.g. by heat) to release the extra stored chemical energy by reverting to the initial reactants. This cyclic process is the least wasteful of on-site resources.

i.e.
$$R \xrightarrow{hv} P \longrightarrow R + energy \qquad (6.6)$$
$$\underset{endergonic}{} \quad \underset{exergonic}{}$$

P is a storable chemical fuel provided that the back-reaction is slow, or alternatively provided that more than one product is formed and these are separable. A further essential is that the energy threshold for photon absorption is low, and the quantum yield high, with no side reactions. At present no system satisfies all of these criteria, but several hold promise of improved efficiency with further development.

The problem of radicals in the reactions above has been avoided by the use of metal-hydrido complexes of the form ML_nH_2, where L is an electron-donating ligand (i.e. a reducing ligand), and M is a transition metal such as cobalt. On irradiation in aqueous solution, these complexes can decompose to give hydrogen, whilst the remaining part of the complex oxidizes the water to hydrogen and oxygen, according to the following series of possible reactions:

$$cis\text{-}ML_nH_2{}^{z+} \rightarrow ML_n{}^{z+} + H_2 \qquad (6.7)$$

$$ML_n{}^{z+} + H_2O \rightarrow ML_n{}^{(z-2)} + 2H^+ + \tfrac{1}{2}O_2 \qquad (6.8)$$

$$ML_n{}^{(z-2)} + 2H^+ \rightarrow cis\text{-}ML_nH_2{}^{z+} \qquad (6.9)$$

The net reaction is type C3:

$$H_2O \xrightarrow{hv} H_2 + \tfrac{1}{2}O_2 \qquad (6.10)$$

This set of equations illustrates very well the transfer of two electrons for the absorption of one photon. The problem is to find a complex which allows the first stage to occur endergonically.

If more than one of this series of reactions is photon assisted, the threshold will be further lowered from the 3 eV level of type C3. For instance, a C4 system has two photons absorbed to halve the threshold energy. This would probably proceed via metastable intermediates, rather than by simultaneous double photon absorption (a low probability process except at laser intensities). The many suggested photochemical schemes for the dissociation of water must all be based on the same energy level diagram: although the route might be different, the threshold energy is always the same.

Alternative schemes for photochemical fuel production suffer from these same limitations: the general need for ultraviolet radiation for high conversion efficiency, the back-reaction rate, the sensitivity to impurities, and wasteful side-reactions. Some possible fuels include methane (from CO_2 and water), methanol and hydrogen peroxide (from oxygen and water in the presence of $KMnO_4$).

A more promising process uses a ruthenium complex, not unlike chlorophyll in structure, to dissociate water, but is still being developed and is not yet well understood. The Ru atom is at the centre of ring structures which will absorb radiation of 377 or 440 nm and which fluoresce deep red (660 nm) even in aqueous solution. When these molecules are modified by the addition of long tails (figure 6.3) they become insoluble in water and no longer fluoresce, the absorbed energy instead being passed to the water, which is dissociated. If a glass plate in water is coated with a mono-molecular layer of the complex, the molecules align with the tails outwards, and

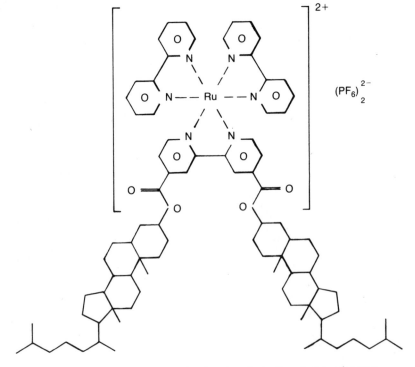

Figure 6.3. Ruthenium complex for the photodissociation of water.

on illumination a gaseous mixture of oxygen and hydrogen is evolved. There are no intermediate radicals, and the quantum yield is about 10%. It may be possible to scale-up the laboratory experiments by stacking plates in water, which will ensure more efficient solar absorption. The expense of this catalyst is only justified if the lifetime is proven to be long, or if revitalizing is easy.

Longer-term solutions involve molecular systems even more akin to biological processes, and may be photochemical or photoelectrochemical in operation. These are based on membranes similar to those which are the site of photosynthesis in plants. The light-absorbing chlorophyll complexes give oxidation/reduction products, which are most likely separated by a membrane, avoiding the back-reactions. Artificial membranes have been made despite the experimental difficulties of handling films less than 10 nm thick. These are either membrane sheets, or small membrane spheres, and can generate a small photovoltage when illuminated.

6.3. *Photoelectrochemical conversion*

If an electrode immersed in an electrolyte is illuminated with radiation of the appropriate wavelength, then a voltage can be set up with respect to a counter-electrode, and a current can be drawn through an external circuit. The first photoelectric effect to be discovered was of this type: in 1839 E. Becquerel reported that a current passed through an external circuit between two similar electrodes in an electrolyte when one was illuminated. The fact that this photoelectric effect has yet to become of technological importance, whilst effects later discovered are the basis of a wide range of successful devices, shows how intractable the problems can be.

Becquerel's discovery has been called the first *photovoltaic* effect, but we shall distinguish between situations in which the absorber is chemically changed by illumination, and those in which electron movement is produced. Both can take place in a bulk medium or at an electrode surface, and both give electrical power without an intermediate thermal process. Photogalvanic cells may store the electrical energy as chemical energy in the same way as conventional 'batteries'.

An electrochemical cell at equilibrium will have two balanced reactions at the cathode and anode (figure 6.4). These may be termed a redox couple, in which an oxidizer, O, accepts electrons from the cathode, and a reducer, R, donates electrons to the anode. There is then an electron flow through the external circuit. The concentration of R will equal the concentration of O unless an overpotential is

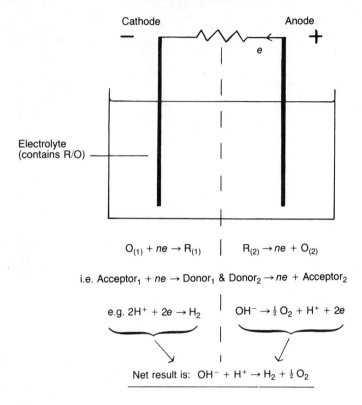

Figure 6.4. An aqueous electrochemical cell, showing the two redox reactions at the electrodes.

applied to the electrodes: this would drive the balance towards either producing O, or consuming O, according to its sign.

We are more interested in upsetting the balance by radiant energy, to produce an electron flow. As in photochemical cells, the back-reaction must be slow. In addition, the electrolyte should be non-corrosive and long-lived, and the internal resistance of the cell should be low. Electrodes which are suitable solar absorbers often decompose rapidly in the electrolyte, whilst others have been found to have a low efficiency for the transfer of electrons to and from the solution.

Water has many of the requirements for a photogalvanic cell, but does not behave as simply as the equation in figure 6.4 suggests (for instance, the existence of H^+ in solution is an oversimplification). In a photogalvanic cell the two electrons needed for the decomposition of one water molecule would be provided by photo-injection from the

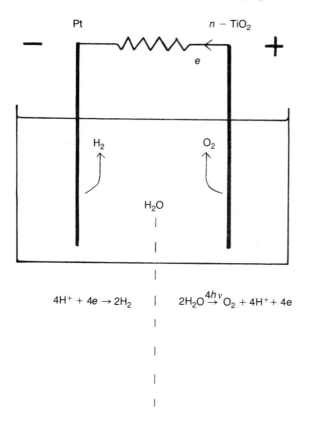

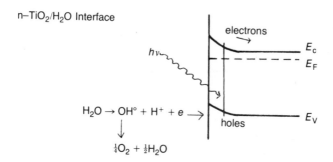

Figure 6.5. A photogalvanic cell for dissociating water.

cathode, or photo-excitation by the anode, and ideally by both photon processes simultaneously:

At anode: $\qquad$ $H_2O \xrightarrow{h\nu} OH^\circ + H^+ + e$ $\qquad$ (6.11)

At cathode: $\qquad$ $H_2O + e \xrightarrow{h\nu} OH^- + H^\circ$ $\qquad$ (6.12)

The creation of free radicals, OH° and H°, will lead to the evolution of O_2 and H_2 at anode and cathode respectively, only if these are of the correct materials.

One semiconductor electrode which has attracted a great deal of attention is n-type TiO_2, for this is stable in an aqueous electrolyte although it has rather a high energy gap (3.5 eV) for efficient solar energy conversion. An anode of this material (either single crystal or polycrystalline) immersed in water will evolve oxygen, with hydrogen evolved at a platinum counter-electrode (figure 6.5). This is a four-electron process, with no wasteful side reactions. Unless the electrolytes at the two electrodes are of different composition, hydrogen evolution may be restricted, but it is aided by the addition of Fe^{3+} ions which assist in electron injection from the cathode. Photons with energy greater than 3.5 eV create free electrons in the TiO_2 and these are removed from the surface layer by the field induced by the surface space charge, as in normal semiconductor junctions. The electrolyte loses electrons to the TiO_2 at this interface and these then pass into the external circuit. The practical realization of this cell must ensure high absorption of solar radiation in the TiO_2 if the quantum efficiency of 10% is to lead to a high solar conversion efficiency. Excess oxygen in the solution can increase back-reactions to an intolerable level.

Another suitable reaction involves the iron–thionine system; thionine is a purple dye and is a more efficient solar absorber than water. This system converts solar energy to electrical energy, rather than to a store of chemical energy, for separation of the products is not practicable. Again we have an endergonic reaction, with few competing side-reactions. The principal reactions may be as follows:

$$TH^+ \xrightarrow{h\nu} TH^+(\text{excited}) \qquad (6.13)$$

$$\underset{\text{thionine}}{TH^+(\text{excited})} + H^+ + Fe^{2+} \rightleftharpoons \underset{\text{semithionine}}{{}^\circ TH_2^+ + Fe^{3+}} \qquad (6.14)$$

$$^\circ TH_2^+ + {}^\circ TH_2^+ \rightleftharpoons TH^+ + \underset{\text{leucothionine}}{TH_3^+} \qquad (6.15)$$

$$Fe^{3+} + TH_3^+ \rightleftharpoons {}^\circ TH_2^+ + H^+ + Fe^{2+} \qquad (6.16)$$

A photoexcited state (threshold about 2·1 eV) in thionine gives up energy to accept an electron from Fe^{2+} by reaction with H^+ ions, forming semithionine (reaction (6.14)). The 'dismutation reaction' (reaction (6.15)) removes semithionine by conversion to leucothionine, and the cycle is completed by a return to semithionine and Fe^{2+} (reaction (6.16)). These form a closed cycle with no net change, except that in sunlight the purple thionine changes to colourless leucothionine, which is reversible in the dark in a matter of seconds. Our interest is not in this photochromic change but in what occurs when two electrodes are immersed in the solution. If only one of these is illuminated by 2·1 eV photons a potential of 100–200 mV is developed across the cell in the following process:

$$^{\circ}TH_2^+ \rightarrow TH^+ + H^+ + e \quad \text{(illuminated electrode)} \quad (6.17)$$

$$Fe^{3+} + e \rightarrow Fe^{2+} \quad \text{(dark electrode)} \quad (6.18)$$

The net effect is the opposite of that given by the photochemical reactions above. Reaction (6.15) is undesirable because it removes the active electrochemical ingredient, semithionine. With solvents other than water it may be possible to suppress this step.

In an experimental cell with a semiconducting electrode of tin oxide-coated glass and a counter-electrode of platinum, it was found possible to illuminate *both* electrodes and still obtain 200 mV from the open-circuited cell or 100 mA m^{-2} from the short-circuited cell (in AM1 solar intensity). It is the surface barrier which is so important with semiconducting electrodes; tin oxide and titanium oxide, are n-type materials and will *accept* an electron from the ions in solution, but will not easily *inject* electrons under the low levels of illumination of solar energy converters.

Despite the low power-conversion efficiency of the iron/thionine cell (0·001%) it is thought to be capable of improvement without changing the basic reactions. The quantum efficiency is satisfactory (62%) but solar absorption is low, for there is only a low percentage of the solar spectrum which is accepted. A mixed-dye solution could cover more of the solar spectrum if chemical interaction can be avoided. Optical absorption of the solutions can only be improved by cunning geometrical arrangements of the cells and electrodes, such as stacks of thin cells.

6.4. Conclusion

Photochemical energy conversion is at a fundamental stage of development, but has the attraction of combined collection and storage of energy without the need for thermal insulation. Either a chemical fuel (such as hydrogen or methane) or electrical energy can be the end product, and in future it may even be feasible to drive certain industrial chemical processes directly by photochemical reactions using solar radiation. This illustrates the need to examine the use to which collected solar energy is to be put: if we require heat, then solar *electrical* generation is not the most logical route, and if we require a chemical product then perhaps it is not necessary to prepare this by a roundabout thermal or electrical method.

7. Direct conversion of solar energy to electrical energy

7.1. *Introduction*

The photovoltaic effect is a technique for utilizing the high grade quality of solar photons, in which the mean solar photon energy of 1·5 eV is converted to 1·5 eV energy rather than to a few $k_B T$ of thermal energy. ($k_B T$ at room temperature has the value 0·025 eV). The maximum efficiency of a single-quantum device is about 40%, for only photons with energy greater than $h\nu_c$ are absorbed, and of these only $h\nu_c$ from each is converted to electrical energy, the excess producing heat. Photovoltaic devices are the most advanced of all quantum energy converters, and provide the most promising path to solar electrical power generation. First we shall take a brief look at two other techniques for the direct conversion of solar energy, although these are disappointingly inefficient for large-scale terrestrial application, even after several years of development work throughout the world.

7.2. *Thermoelectric converters*

Only in semiconductors are thermoelectric effects strong enough to use in solar energy conversion. The generation of an e.m.f. in a thermocouple, most evident as a potential difference, V, across an open circuit in the couple, depends on the temperature difference according to the Seebeck coefficient, S, for the pair of materials used.

$$S = \frac{dV}{dT} \tag{7.1}$$

The value of S for a metal thermocouple is small, but for a semiconductor thermocouple it may be several hundred $\mu V\ K^{-1}$ and several couples in series/shunt combinations can make a usable battery.

For optimum efficiency the resistance (R_L) of a load on a thermoelectric generator should be related to the internal resistance (R_{int}) by the following:

$$R_L = M R_{int} \tag{7.2}$$

This gives a maximum efficiency of:

$$\eta_{max} \doteq \left(1 - \frac{T_{cold}}{T_{hot}}\right) \frac{(M - 1)}{[M + (T_{cold}/T_{hot})]} \tag{7.3}$$

Not surprisingly, in what is after all just a heat engine operating between a source at temperature T_{hot}, and a sink at temperature T_{cold}, high efficiencies require large temperature differences, and hence solar concentrators. The expression (equation (7.3)) is the Carnot efficiency multiplied by a factor which includes the influence of irreversible heat conduction and Joule losses. The factor M depends on a quantity called the thermoelectric figure of merit (Z) and the mean temperature.

$$M = \left(1 + Z\frac{(T_{hot} + T_{cold})}{2}\right)^{\frac{1}{2}} \tag{7.4}$$

In a couple composed of n-type and p-type pieces of the same semiconductor, having electrical conductivity σ and thermal conductivity κ:

$$Z = S^2 \sigma/\kappa \tag{7.5}$$

The need for a good electrical—but poor thermal—conductor is easily understood: the couple must not lose energy through I^2R heating and also should not act as a thermal short-circuit. For semiconducting compounds based on the elements Bi, Pb, Sb with Te and Se, the value of Z is about $0·003$ K^{-1}.

Since Z decreases with increasing temperature, it is best not to operate a single couple across a large temperature difference. Instead the high-efficiency limit set by high temperatures can be approached by 'cascading' several couples in series between the source and sink. Each couple is chosen so as to operate at its optimum Z value, and the cold junction of one couple is thus at the same temperature as the hot junction of the next in line (figure 7.1). The stacking of several semiconducting couples in this way introduces problems of thermal expansion mismatch, which may be reduced by substituting sintered slabs for single crystals, at the expense of a reduced Z value.

Limiting practical efficiencies of 16% have been calculated for advanced designs, but 5–8% is a good experimental achievement. The USA 'SNAP' generators for satellites, which operate from radioactive sources, have an efficiency of $5·5\%$. Unless new semicon-

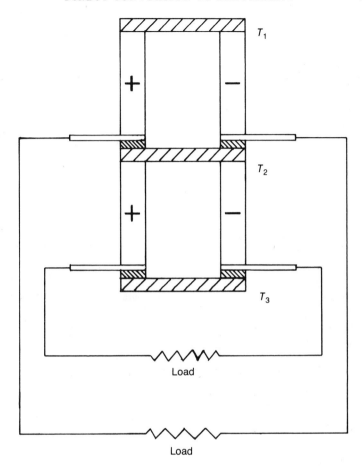

Figure 7.1. Cascaded thermoelectric couples. Here $T_1 > T_2 > T_3$.

ducting materials with substantially higher figures of merit can be found, this solar power-conversion route is not very attractive for any but the most specialized application.

7.3. *Thermionic converters*

The use of thermionic emission as a thermal-to-electrical power converter was suggested in 1915, but has yet to be applied on a noticeable scale. Photoemissive devices, using solar-photon absorption in a material to give an electron emission current into a vacuum, have too low a quantum efficiency to be useful converters, but by changing the photons to thermal energy for heating the electron emitter, it is possible to obtain a more useful electron flux.

The current density from a thermionic emitter is given by the Richardson equation:

$$J_k = AT_k^2 \exp(-\varphi_k/k_B T_k) \tag{7.6}$$

where A = the Richardson constant ($1 \cdot 2 \times 10^6$ A m^{-2} K^{-2} for tungsten), T_k = emitter (cathode) temperature in K, φ_k = work function of the emitter surface, k_B = Boltzmann constant ($8 \cdot 62 \times 10^{-5}$ eV K^{-1} = $1 \cdot 38 \times 10^{-23}$ J K^{-1}). For most metals the work function lies between 3 and 5 eV but by special treatments this may be reduced even to negative values. The commonly used caesium-coated tungsten cathode has a work function of about 2 eV, and an A value of about 3×10^4 A m^{-2} K^{-2}. Equation (7.6) suggests that we require a large value of T_k and a small value of φ_k for high electron emission density, and this is shown more clearly by the graph in figure 7.2. Appreciable emission ($>10^4$ A m^{-2}) is obtained from the Cs–W cathode only above 1250 K, and this definitely means that concentrator systems must be used.

Assuming that we have managed to boil off some electrons, we should like to pass them through an external load. This is accomplished by adding close to the cathode, an anode with a sufficiently low work-function (φ_a) that the collected electrons will be at different potential energy from their potential energy when in the cathode). Figure 7.3 is a potential energy diagram of this arrangement with the vacuum level and Fermi energy level shown for each electrode. The difference in the work function ($\varphi_k - \varphi_a$), therefore, generates the e.m.f. of the device, and this is unfortunately low.

The reason for the narrow separation of the anode and cathode is the electron space charge, but placing the two metal plates less than $0 \cdot 5$ mm apart, when one of them is to be heated to more than 1000° C, invites a short-circuit as thermal expansion occurs. A little caesium vapour in the interspace allows a larger plate separation, for when the caesium is heated by the cathode it is readily ionized, and neutralizes the negative space charge.

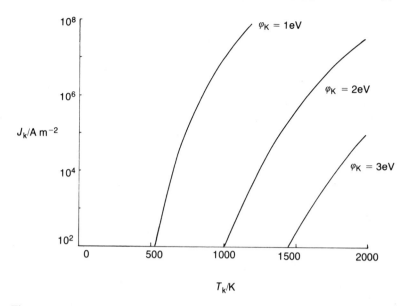

Figure 7.2. Thermionic emission from cathodes with various work functions, according to equation (7.6).

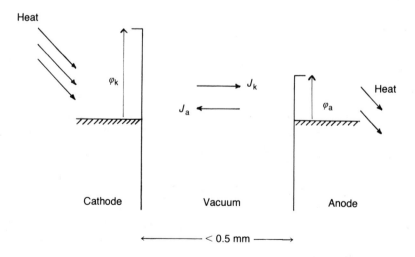

Figure 7.3. Energy levels of a thermionic converter. A current density of $(J_k - J_a) = (\varphi_k - \varphi_a)/R_L$ would flow through a resistive load, R_L.

The anode will be heated by the kinetic energy of the electrons, and further heat reaches it by radiant emission from the cathode. If the temperature rises above 500° C electron emission may take place from the anode and oppose the required cathode emission. The need to cool the anode can represent a loss of input energy unless the cooling fluid is treated as an energy supply. Extra heat losses are incurred by conduction in the cathode supports.

A device operating with $T_k = 1500$ K and $T_a = 750$ K would still have a Carnot efficiency of 50%. In practice less than 20% is achieved. In terms of actual power delivered to a load this would be $5–40 \times 10^4$ W per m^2 of cathode area.

Realistic assessment of the device must include the expected operating lifetime, and this is only a few thousand hours (i.e. months rather than years). Because of this, and the expense of the device, applications have been limited, although both radioactive and solar sources have been used.

7.4. *Photovoltaic converters*

The photovoltaic effect has been called an *internal* photoemission process. It differs significantly from photoemission in having a lower threshold for photon absorption, but is similar in its direct conversion of light to electricity without a thermal stage. Similarities between wet photogalvanic cells and dry photovoltaic cells are several, but it is only the photovoltaic cell at present which has sufficiently high optical absorption and sufficiently low electrical resistance for the conversion of solar energy on a useful scale. Photovoltaic cells are the only direct solar converters likely to become cheap enough for economic electrical power generation.

There is apparently a wide choice of suitable semiconductors with the appropriate spectral absorption range, so that in theory we can select a material which will match the solar spectrum. In practice the choice of materials is limited by various conflicting requirements. The photovoltaic effect is the simultaneous generation of an e.m.f. in the illuminated cell and the passage of a current through a load connected across it. Early semiconducting materials such as Cu_2O and Se showed this property weakly, but were of little use for power generation. Since 1954 when the Bell Telephone Laboratories in the USA announced a photovoltaic battery using silicon p–n junctions, this effect has been taken very seriously as a means of direct solar energy conversion.

All present solar cells have three features in common:

(a) an optical absorber which converts photons to electron-hole pairs;
(b) an internal electric field which separates these charges before they can recombine;
(c) contacts at the ends of the semiconductor to make connection with an external load.

These are not necessary physically separated within the cell. For instance, the most effective optical absorption will take place *within* the electric field region.

The photon absorber is, of course, the semiconductor itself, which is chosen to have a bandgap matched to the solar spectrum. It might be thought that a *low* value of this gap, E_g, would be ideal since this would allow the semiconductor to absorb almost the whole range of solar radiation, except a small part of the near infra-red. But the e.m.f. of a cell is limited by E_g, and if this is small then the cell e.m.f. will be

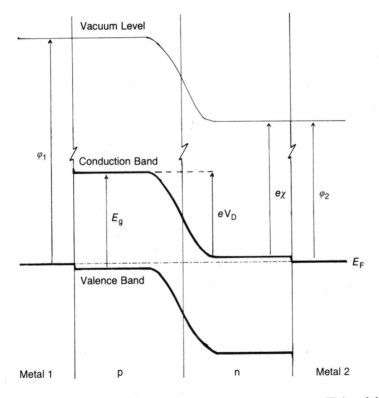

Figure 7.4. Energy levels in a p–n junction. ($e\chi$ is the electron affinity of the semiconductor, and φ_1, φ_2 are the metal work functions.)

even smaller. The energy level diagram for a p–n junction (figure 7.4) shows that at most we can expect the Fermi levels to coincide with the appropriate band edges, giving a maximum band bending of E_g under open-circuit conditions, and hence a maximum open-circuit voltage limited by E_g. Also, it is known that the probability of creating *two* or more pairs of electrons and holes from photon of energy greater than *twice* E_g is low, unless the incident intensity reaches a very high level. Thus, we expect at most only one electron-hole pair per absorbed photon, and the excess energy $(hv - E_g)$ is dissipated by phonons in the crystal lattice. We can either collect a large number of 'low voltage' photons which provide much current, or we can collect a small number of 'high voltage' photons which provide a small current. For maximum *conversion efficiency* of the cell, a value of E_g between these extremes must be selected, according to the actual distribution of solar photons.

The way in which the efficiency depends on E_g is shown in figure 7.5. It has a peak which becomes less pronounced in weak sunshine

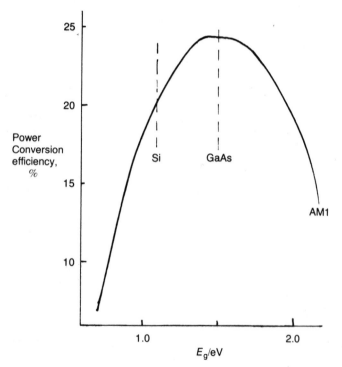

Figure 7.5. Theoretical maximum efficiency of solar cells as a function of their bandgaps, E_g.

since the infra-red end of the spectrum is most attenuated by the atmosphere. Details of the spectrum such as the absorption bands of water, carbon dioxide and other atmospheric constituents are not too important since they mainly affect this weak infra-red end, beyond the response of most semiconductors likely to be used for the purpose. This curve suggests that $1\cdot4$–$1\cdot6$ eV is the best range for the bandgap of semiconductors to be used in solar cells. We have also to choose between a direct gap semiconductor such as GaAs, and an indirect gap semiconductor such as Si. Leaving aside considerations of the way the materials are prepared and the present state of their technology, we might expect to use a direct gap material, with its higher absorption coefficient (figure 7.6). This would mean that thinner slices would suffice to absorb photons of a given energy. Notice that although the *indirect* gap material starts in this case to absorb photons of energy lower than the *direct* gap material, the direct gap material soon overtakes this and may give a better response over the solar spectrum range.

A second look at the physical process suggests that this may not be an advantage at all. If absorption is strong, then most of the photons will be converted to electron–hole pairs within a short distance of the surface. If we can arrange the separating field to lie within this volume, then the carriers may be separated rapidly, but this is difficult in practice with the usual p–n junction device. Unless we remove the electrons from the vicinity of the holes (or vice versa) they will combine, especially near a defect or the surface of a crystal. (The surface itself is a special example of a defect: it is a break in the regular array of lattice sites.) A local change in the energy of a charge carrier near to a defect tends to hold it there briefly, long enough to attract an oppositely-charged carrier with consequent recombination. Some way of reducing the surface recombination rate is essential if we are to take advantage of the high optical absorption of direct gap materials.

If no potential difference was built into the semiconductor, we should need an external voltage source to circulate the photocurrent through an external load, as with photoconducting devices. We have already seen that one such field can arise at the surface of a semiconductor unless care is taken to prevent it. This surface field is important in the modified version of the solar cell known as a metal-semiconductor or Schottky diode. True surface fields cannot be used in solar cells since we must apply metal contacts to remove the carriers from the semiconductor. Any gross imperfections in the regularity of the crystal lattice will have associated with it an internal

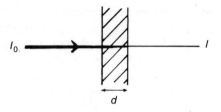

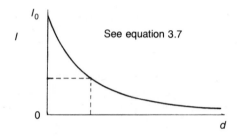

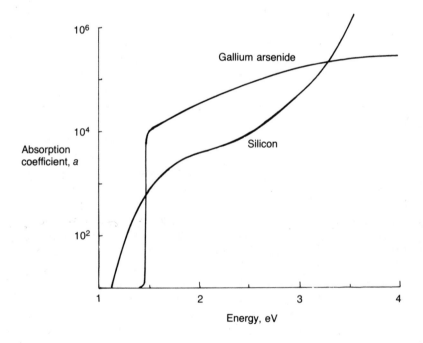

Figure 7.6. Optical absorption of semiconductors.

electric field which can, in principle, be used to generate a photovoltage. The p–n junction between oppositely doped regions of a semiconductor is the most important for present commercial cells, but p–n junctions between *dissimilar* semiconductors with different energy gaps (heterojunctions) are used in cells which are being developed. In our p–n solar cell the electrons and holes created within the field region, or within a short distance on either side, are separated by the diffusion potential. The electrons flow 'downhill' towards the n-type side and the holes float 'uphill' towards the p-type side. Electrons and holes may start as minority carriers, but become majority carriers when they have crossed the junction. The M–S, Schottky, diode has a similar energy band diagram and in its simplest form behaves in the same way as a p–n junction (figure 7.7).

If there is no load and the cell is left open-circuited under illumination (as in the figure), then the separation of positive and negative charges by the built-in field produces a potential difference between the contacts which is sufficient to oppose further net charge flow. The formation of ohmic contacts which transfer carriers from the semiconductor to the load is part art and part science, and is more difficult for the larger gap materials. In the simple theory it is the work function of the metal which determines the height of the potential barrier between it and the semiconductor. For the barrier to vanish the work function should be less than the electron affinity of the semiconductor for n-type material, and greater than the electron affinity for p-type material (see figure 7.4). One of the contacts to p–n cells is generally grid-shaped to allow light to pass through that face of the crystal. In a Schottky barrier cell, the corresponding contact is semitransparent and covers the whole face of the crystal; its electrical resistance is high since it is very thin, and a grid contact is again generally needed on top of the metal layer (figure 7.8).

We shall now study the full current–voltage (I–V) characteristic of a solar cell (figure 7.9) to understand how the cell acts as a power source. The 'dark' and 'illuminated' characteristics are of the same shape, but are separated vertically by the photo-generated current, $-I_L$. (In certain cells the two curves cross in the first quadrant instead of running together.) For simple p–n junction cells, the following equation describes these characteristics:

$$I = I_D - I_L \tag{7.7}$$

total,   dark,   illuminated, components of the current

and,

$$I_D = I_0[\exp(eV/k_B T) - 1]. \tag{7.8}$$

Here I_L depends on the illuminated area of the cell, and I_D on the junction area, the difference in these areas being that of the opaque top contact. Under large reverse bias (V negative), equation (7.8) shows that I_D approaches $-I_0$, the reverse saturation (or leakage) current. Under forward bias (V positive), I_D approaches a simple exponential dependence on V.

The maximum photocurrent, I_L, produced within the cell by solar radiation can be calculated from the number of photons of energy greater than the energy gap of the semiconductor which actually enter the semiconductor, are absorbed, and produce charges collected by the contacts.

i.e. $$I_L = N(>E_g) \cdot (1 - p)[1 - \exp(-at)]e \, \eta_{col}, \tag{7.9}$$

where $N(>E_g)$ is the number of solar photons with energy greater than E_g; $(1 - p)$ is the fraction *entering* the semiconductor; $[1 - \exp(-at)]$ is the fraction actually *absorbed* in thickness t; and η_{col} is the carrier collection efficiency. Since a depends on wavelength, this equation should be evaluated at each energy in the solar spectrum above E_g. The reflection loss, p, can be estimated from the refractive index of the semiconductor (equation (3.5)) which, in turn, can be estimated by an empirical equation involving the energy gap.

$$n = (173/E_g)^{\frac{1}{2}} \quad (E_g \text{ measured in eV}). \tag{7.10}$$

For silicon, with refractive index 3·5, only about 69% of the incident photons enter the cell unless an anti-reflection coating is added. These calculations suggest a maximum value of I_L of about 40 mA cm^{-2} in silicon p–n cells under AM1 illumination.

When the photovoltaic diode is used as a source of power, the fourth quadrant of the characteristic is often shown inverted for convenience. Figure 7.10 uses this convention together with a plot of the current drawn by a load, R_L, to show the operating point of a cell under load. For the extraction of maximum power from any source load impedance must equal the source impedance. For the solar cell this operating point of maximum efficiency (P_{max}) can be found approximately by constructing the diagonal of the rectangle which passes through I_{sc} and V_{oc}, as shown, and finding its intercept with the cell characteristic. The gradient of the diagonal closely gives the matched load resistance.

A further measure of cell behaviour is the fill factor (FF) which is

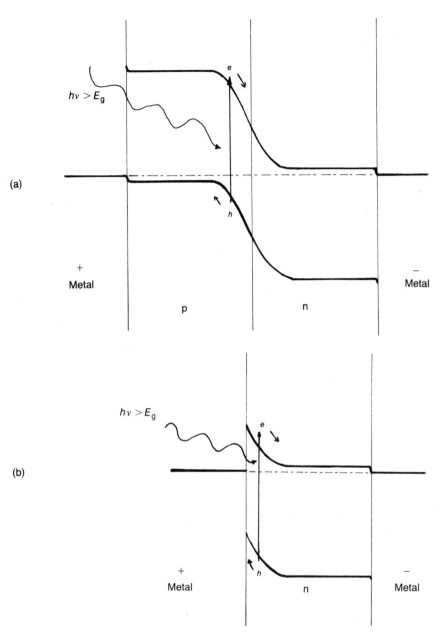

Figure 7.7. Charge carrier separation within solar cells: (a) p–n junction; (b) Schottky junction. e are electrons, h are holes.

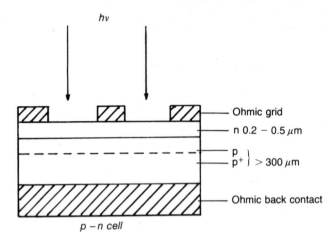

p – n cell

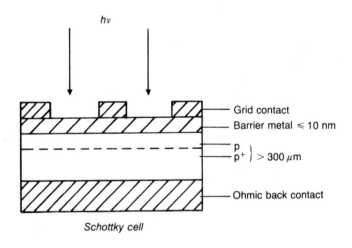

Schottky cell

Figure 7.8. Schematic illustration of cross-sections through solar cells.

the ratio of the rectangular area defined by $I_{mp} \times V_{mp}$, to the rectangular area defined by $I_{sc} \times V_{oc}$, i.e.

$$FF = (I_{mp} \cdot V_{mp})/(I_{sc} \cdot V_{oc}). \qquad (7.11)$$

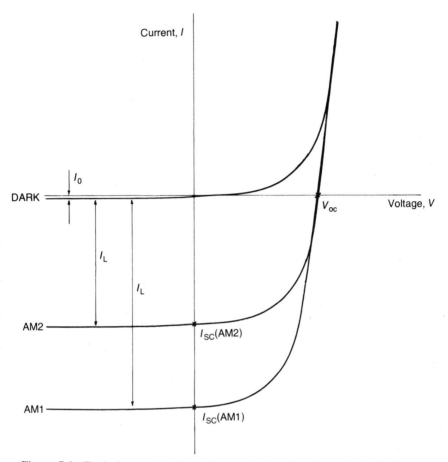

Figure 7.9. Typical current-voltage characteristics of a photovoltaic diode, dark and with irradiances AM1 and AM2.

This is a measure of the squareness of the *I–V* characteristic, or of the reduction in voltage and current at the P_{max} point compared with the open-circuit and short-circuit conditions. Ideally we should like FF to approach unity, but by the very nature of a diode it is limited to less than 80%. In an ideal cell which has no series or shunt resistance losses, the fill factor is more accurately called the curve factor (CF). The CF is still limited by the exponential nature of the electrical conduction through a diode to less than unity.

The diode dark current (equation (7.8)) is often more closely given by:

$$I_D = I_{01}[\exp(eV/nk_BT)-1], \text{ where } n > 1. \qquad (7.12)$$

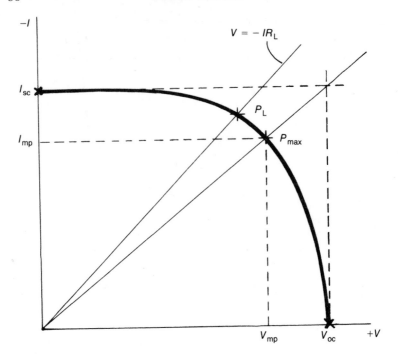

Figure 7.10. Fourth quadrant characteristic of a solar cell. P_L is the operating point of the cell when a load R_L is connected. P_{max} is the operating point for maximum efficiency, requiring a load V_{mp}/I_{mp}.

In these cells the CF is reduced from its value when $n = 1$, by additional routes for the transport of carriers across the junction. The dark current in a real p–n solar cell should be written as the sum of several terms:

$$I_D = I_{01}[\exp(eV/k_BT)-1] + I_{02}[\exp(eV/2k_BT)-1]+ \ldots (7.13)$$

I_{01} is the usual diffusion current given by equation (3.9) and for good quality silicon is about 10^{-7}–10^{-8} A m^{-2} at room temperature. I_{02} is a term for the recombination of electron–hole pairs at a single defect energy level within the depletion region (width, W):

$$I_{02} = eWn_i/(\tau_p\tau_n)^{\frac{1}{2}}. \qquad (7.14)$$

Further leakage current terms allow for tunnelling of carriers across the junction (particularly in highly doped diodes), or for recombination at a series of energy levels distributed across the energy gap.

Under normal diode operation, recombination currents are the most important at low bias levels, and diffusion current at higher bias levels. The value of n in equation (7.12) thus increases with forward bias. Unfortunately for simplicity, a solar cell is usually operated near the knee of the I–V curve, at which point *both* leakage currents are of significant value. This explains why a single dark-current term is widely used, and why a non-integer value of n is accepted.

Metal–semiconductor diodes have a simpler expression for the principal leakage current which is given by applying thermionic emission theory, that is, the Richardson equation applies.

$$I_o = A^* \cdot T^2 \cdot \exp(-\varphi_B/k_B T). \tag{7.15}$$

Typically, I_o is 10^3 times more than the value for p–n junctions in the same material, and is $3 \cdot 6 \times 10^{-4}$ A m^{-2} for p-type silicon if the barrier height (φ_B) between metal and silicon is $0 \cdot 8$ eV. A^* is the effective Richardson constant for the semiconductor, and differs for p- and n-types. It is defined in the same way as the constant A for thermionic emission from metals but uses an effective mass which differs from the free electron mass.

We obviously require all components of the dark current, whatever the diode type, to be as low as possible for an effective solar cell, since I_D flows in opposition to the light-generated current.

From equations (7.7) and (7.8) for the ideal photovoltaic diode, the open circuit voltage is given by:

$$I_L = I_o[\exp(eV_{oc}/k_B T) - 1]. \tag{7.16}$$

We can obtain an explicit relation for V_{oc} by turning this around:

$$V_{oc} = \frac{k_B T}{e} \ln \left(\frac{I_L}{I_0} + 1 \right) \tag{7.17}$$

Usually I_L will be much greater than I_o and the integer can then be omitted. If we have a cell in which the ideality factor, n, is greater than unity, then V_{oc} is increased by this factor. This may not, in practice, lead to a better cell because increases in n are accompanied by compensating increases in I_o, and a high V_{oc} requires a low I_o. The ratio of eV_{oc} to E_g is the voltage factor (VF), a measure of how closely the limiting V_{oc} is approached.

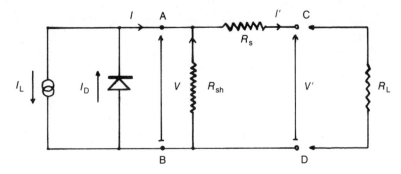

Figure 7.11. Equivalent circuit of a solar cell.

The value of the short-circuit current can be found in a similar manner: by letting the output voltage be zero in equations (7.7) and (7.8).

$$I_{sc} = -I_L \qquad (7.18)$$

In a real device, the series and shunt resistances may not have the respective low and high values desired. These lead to losses in output, which may be readily understood from an examination of the equivalent circuit of the cell. Figure 7.11 shows an electrical circuit consisting of discrete components connected in such a way that it has the same characteristics as many solar cells. Of course, this is not quite true, since a solar cell does not have its resistive parts physically separated from the diode junction and from the current generator, but this equivalent circuit will behave in much the same way as an actual cell to any external influence such as an applied bias.

To the left of the points A and B the circuit elements are those of our ideal cell, with behaviour described by equations (8) and (9). The addition of series (R_s) and shunt (R_{sh}) resistances to the cell is represented by the additional elements to the right of A and B. These have been added as single 'lumped' components, but may really be distributed in a three-dimensional array throughout the cell, for which a more complicated resistor network would be needed in our model. Their origin does not concern us here, but will be explained in the next chapter.

The output voltage (V') and current (I') at C and D, available for connection to an external load are now given by the following equations.

$$I' = I_D - I_L + \left(\frac{V' - I'R_s}{R_{sh}} \right) \qquad (7.19)$$

and

$$I_D = I_0\{\exp[e(V' - I'R_s)/k_BT] - 1\}. \tag{7.20}$$

The term concluding equation (7.19), $I'R_s/R_{sh}$, may be omitted, since R_{sh} is very much larger than R_s. These losses have a significant effect on the V_{oc}, I_{sc} and FF: the fill factor is affected by both resistance elements, but a poor series resistance affects only the I_{sc} value and a poor shunt resistance affects only the V_{oc} (figure 7.12).

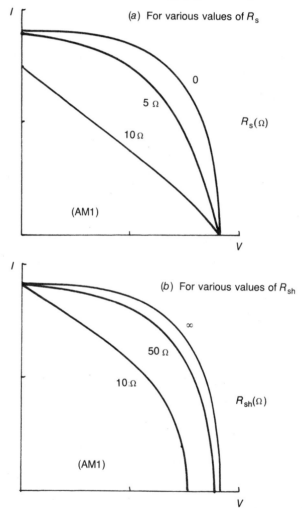

Figure 7.12. The effect of resistance losses on p–n Si solar cell characteristics (cell area 2 cm²).

The only circuit element appearing in figure 7.11 which has not been discussed is the external load, R_L. We have said that the cell will operate at its maximum efficiency only if the load is matched to the effective resistance at the P_{max} point. Since this resistance, R_{mp}, depends on the leakage current (through V_{mp}) and hence ultimately on the bandgap of the semiconductor, wider gap materials than silicon (such as gallium arsenide) have a larger R_{mp}, and so can tolerate a greater series resistance with a given load, R_L, before the FF is degraded.

We have seen that the photovoltaic effect is capable of producing high conversion efficiencies for electricity from solar radiation, if cells can be made which behave according to theory. Even with certain avoidable losses within the cell, we shall see that useful efficiencies are achieved by production devices. It is not the technique which is in question, but the economics which prevent large-scale use. Practical attempts to improve the performance of standard p-n cells will be discussed in a later chapter, and will be contrasted with the alternative approach of developing less expensive, but less efficient, cells.

8. Flat-plate thermal collectors

8.1. *The basic collector*

Flat-plate solar collectors are the easiest to analyse, as well as the simplest to construct. They have been installed in hundreds in Australia, Israel, Japan and the USA, and are appearing in the UK. There now exists a large quantity of operating data for solar water heaters in various climates and latitudes, although the correlation of performance data from different sources is made difficult by incomplete details in some reports, and by different energy extraction techniques in others. We shall attempt to distil the common factors from this situation.

The collector pictured in figure 8.1 is the simplest workable

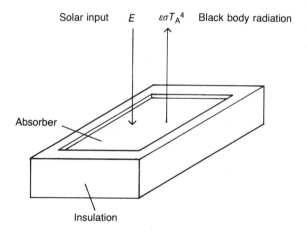

Solar input E $\varepsilon\sigma T_A^4$ Black body radiation

Absorber

Insulation

Figure 8.1. A simple flat-plate solar collector of unit area in equilibrium.

structure. It is a blackened absorbing plate facing south, and insulated from its surroundings except over the top face. For the present we shall leave aside the extraction of heat from the absorber, and omit

any details of the associated plumbing. If the integrated solar input is
E, the plate will absorb aE and heat up to a temperature T_A.
Simultaneously, it will radiate to the atmosphere. If there were no
additional energy losses or gains, for unit area at equilibrium:

$$aE = \sigma \varepsilon T_A{}^4. \tag{8.1}$$

For a black body absorber in AM2 sunlight, T_A would be 345 K
which happens to be about the temperature reached by an insulated
black plate left in the sun, even though it is open to other paths for
heat loss and gain.

There is an additional convection loss of about 4 W m^{-2} in still air
for each 1 K excess over the ambient temperature; on a windy day
this would be several times greater. In the present example, we will
lose around 200 W m^{-2} by still-air convection, and a much smaller
amount by conduction to the atmosphere. Now, this convection loss is
almost exactly balanced by the radiation received from the atmospheric water vapour and CO_2, at wavelengths beyond 5000 nm. (In
humid weather this may amount to as much as 300 W m^{-2}.) If we
were to treat the atmosphere as a black-body radiator at 300 K (i.e.
take the 'sky temperature' to be the same as the air temperature)
then the sky radiation would be 460 W m^{-2}, which suggests a value
for the atmospheric emittance of 0·65, ignoring the spectral distribution of the atmospheric radiation. This longwave emission is important at night since it helps to prevent the Earth's surface from cooling
by radiation. Our energy balance has also omitted thermal exchange
with nearby objects, or the ground, but these are not easily general-
ized: most heat-flow texts deal only with symmetrical objects, and not
with trees or roof-tops.

There are many obvious improvements to be made to this simple
structure before it is a useful device, although certain additions to the
collector may be uneconomic compared with the monetary and
energy costs of the improvement. The lifetime of the collector enters
these considerations, for it is the total energy output from the device
which must be balanced against the energy used to produce it. Let us
first consider the addition of a transparent cover. A large sheet of
glass is an expensive and fragile item, but the alternative plastics show
poor long-term behaviour when exposed continuously to ultraviolet
light. The spacing of this cover from the absorber must be sufficient
to leave room for two insulating stagnant films of air (on the inner
faces of the cover and absorber). The cover will then reduce
convection losses from the absorber by settling at an equilibrium
temperature between that of absorber and ambient air.

Unless the iron content of a glass is high, it will transmit at least 85% of visible solar energy at near-normal angles of incidence. The addition of a good anti-reflection coating will give broad-band transmittance of over 90%, with an absorptance loss of 5% (see figure 8.2). We shall assume that the temperature gradient across the

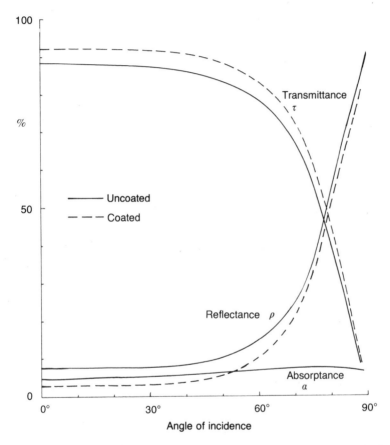

Figure 8.2. The AM2 reflectance, transmittance and absorptance of a single sheet of glass, before and after coating both surfaces with an anti-reflection layer. (Coating and figures prepared by OCLI, California, USA.)

glass plate is negligible compared with the gradient across the stagnant air films. The new equilibrium temperature of the absorber will now be estimated, using the energy balance scheme shown in figure 8.3. A linear conduction/convection coefficient (h) of 5 W m^{-2} K^{-1} will be used.

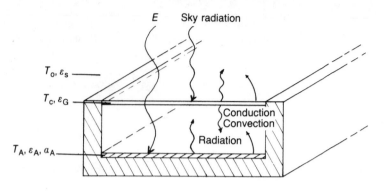

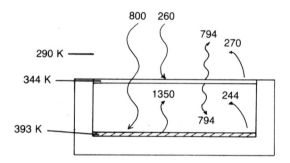

Figure 8.3. Energy balance of a flat-plate solar collector with a glass cover. (Energies in W m^{-2}.)

For the *absorber*, the energy balance per unit area (assuming the cover is transparent to most of the solar spectrum) is as follows:

$$a_A(E + \varepsilon_G \sigma T_c^4) = \varepsilon_A \sigma T_A^4 + h(T_A - T_c)$$

solar input	radiated from cover	radiated from absorber	conduction and convection to cover

(8.2)

A similar equation may be written for the cover sheet:

$$a_G(\varepsilon_A \sigma T_A^4 + \varepsilon_s \sigma T_o^4) + h(T_A - T_C) =$$

radiated from absorber	radiated from sky	conduction and convection to cover

$$h(T_c - T_o) + \varepsilon_G \sigma T_c^4 + \varepsilon_G \sigma T_c^4$$

conduction and convection to sky	radiated to sky	radiated to absorber

(8.3)

where: a_A, ε_A, T_A refer to the absorber; we assume $a_A = \varepsilon_A = 1$; a_G, ε_G, T_c refer to the cover; we assume $a_G = \varepsilon_G = 0$ for visible and $a_G = \varepsilon_G = 1$ for infra-red. $\varepsilon_s = 0.65$, with an effective sky temperature of $T_0 = 290$ K.

It is possible to approach this problem in various ways (e.g. to combine the emissivities of the cover and absorber into one effective emissivity for the collector), but the answers for T_C and T_A will be little affected. From equation (8.3), substituting for T_A terms from equation (8.2), we can find a value for T_C. If the solar input is 800 W m^{-2}, and with the absorptances and emittances given above, T_C is 344 K. This value is put into equation (8.2), to find that T_A is 393 K. Although there has been no heat extraction from the absorber for useful work, the collector must be designed to withstand such a temperature in case the working fluid stagnates.

The relatively high cover temperature is reasonable, and higher than that of most building windows because of the large net radiative (550 W m^{-2}), conductive and convective (240 W m^{-2}) input from the underlying surface. Adding further sheets of glass reduces the convective losses even more, but any benefit is eventually limited by the absorptance of glass, and by internal reflections between the plates. If the cover is placed very close to the absorber it will heat up considerably more than the glass plate in the above situation, because of increased conduction. Indeed, if a slab of infra-red-opaque plastic is placed *on top* of a blackened plate it may be destroyed before the equilibrium temperature is reached, for its thermal conductance is low and yet it is in contact with a hot surface, which is also radiating. Heat trapped in a Perspex slab in contact with a blackened absorber has raised the plastic's temperature to 447 K in sunlight, before destruction, with an estimated stagnation temperature of over 500 K. The Perspex thermal trap collector can be more efficient than simpler flat-plate designs at high absorber temperatures or low sunlight levels since it has a lower thermal loss to maintain and a higher thermal capacity.

8.2. *Achievement of higher temperatures*

It would be useful to eliminate the convective exchange between absorber and cover. Whilst various baffle arrangements and honeycomb structures have been suggested, these can increase the *conduction* path from the absorber plate, and the only effective way to eliminate convection is to remove the air which carries the heat away. Evacuation results in a rather fragile collector if the flat-plate pattern is retained, for the cover glass must withstand atmospheric pressure.

A more practical evacuated collector is a bank of small-diameter tubes, similar in size to fluorescent light fittings. These solar collectors have been constructed by several industrial concerns using existing technology. For the purposes of comparison, we shall examine the absorber and cover temperatures of a flat-plate geometry, and shall find that high plate temperatures are possible (see figure 8.4).

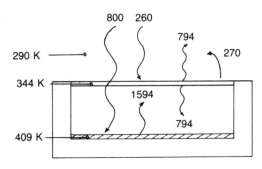

Figure 8.4. Energy balance of an evacuated flat-plate solar collector. (The figures represent power in W m^{-2}.)

For the *absorber*, the energy balance per unit area is as below, since the cover is again transparent to most of the solar input.

$$a_A(E + \varepsilon_G \sigma T_C^4) = \varepsilon_A \sigma T_A^4. \tag{8.4}$$

For the *cover*, the similar balance is given by:

$$a_G(\varepsilon_A \sigma T_A^4 + \varepsilon_s \sigma T_o^4) = \varepsilon_G \sigma T_C^4 + \varepsilon_G \sigma T_C^4 + h(T_C - T_o). \tag{8.5}$$

A solar input of 800 W m^{-2} gives a cover temperature of 344 K, as before, but on substitution into equation (8.4), an increased absorber temperature of 409 K is estimated. Note that although the absorber conduction/convection losses are eliminated in this collector, the radiation loss is higher.

Still higher temperatures should be possible if the absorber has a selective surface. Whilst surfaces with a solar absorptance approaching 0·9 and infra-red emittance approaching 0·1 have been produced, the long-term stability of some of these at their operating temperatures is not always sufficient to justify the rather expensive production techniques. On simple flat-plate collectors, which we have seen can lose almost as much energy by conduction/convection as by radiation, the advantage of selective absorbers is minimal when compared with the increase in actual working temperature. Table 8.1

TABLE 8.1. *Stagnation temperature of solar collectors (K).*

Collector structure	Black-body absorber		Selective absorber	
	T_A	T_C	T_A	T_C
1. 'Naked' absorber—simple theory	345	—	598	—
2. Single transparent cover (Air space)	393	344	536	338
3. Single transparent cover (Vacuum space)	409	344	712	338
4. Single transparent cover (Vacuum space) $+CR=2$	478	393	829	384
5. Single transparent cover (Vacuum space)				
$+CR=10$	714	588	1212	560
$+CR=10^2$	1288	1077	2180	1024
$+CR=10^3$	2302	1933	3987	1883

$(T_o = 290$ K, $\varepsilon_s = 0.65$, $E = 800$ W m^{-2})

$a_G = \varepsilon_G = 1$ Infra-red

$a_G = \varepsilon_G = 0$ Visible

$a_A = \varepsilon_A = 1$ for black-body absorber

$a_A = 0.9$ Visible, $\varepsilon_A = 0.1$ Infra-red, for selective absorber

gives the absorber and cover temperatures for the three collector structures already discussed, but with the addition of a selective coating on the absorber (see figure 8.5). Notice that the cover temperature is slightly lower when a selective coating is used. Practical flat-plate collectors using aluminium absorber plates coated with a selective chromium oxide have been used to heat a water/anti-freeze mixture to 383 K for space heating, and some designs of tubular collector have stagnation temperatures approaching 600 K.

An unfortunate characteristic of nearly all selective surfaces is that a(visible) tends to be less than 0·9, and although ε(infra-red) may be very low this cannot compensate for incomplete solar absorption at the beginning of the chain. It is only at very high collector temperatures that this behaviour can be tolerated. The a/ε ratios of 12–15 quoted previously were for a 2–5 μm layer of silicon on top of a high reflectivity metal substrate (e.g. silver), together with an anti-reflection coating on the outer surface (see figure 8.6). This absorber/reflector combination *will* operate satisfactorily at high

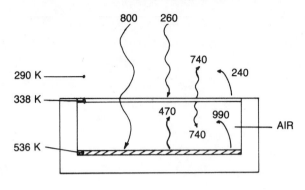

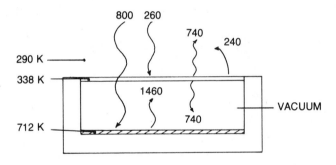

Figure 8.5. Energy balance of a flat-plate solar collector, with an air space or a vacuum space, and a selective coating on the absorber. $(a(\text{visible}) = 0.9,$ $\varepsilon\ (\text{infra-red}) = 0.1.)$ (The figures represent power in W m^{-2}.)

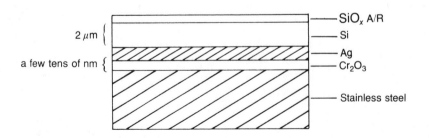

Figure 8.6. Selective absorber developed by B. O. Seraphim.

temperatures without excessive degradation, but it must be protected against the atmosphere. Solar radiation of wavelength shorter than $1.1 \mu m$ will be absorbed in the silicon to generate heat, which is conducted to the underlying metal substrate. This metal will not lose heat rapidly by radiation because of the high infra-red reflectivity of the thin metal film on top. The high operating temperature of this absorber might enhance diffusion between the silicon and the steel, but this is prevented by the thin layer of chromium oxide on the steel surface which closely matches the thermal expansivity of steel and so will not crack on repeated thermal cycling. The stability of thin metallic layers at elevated temperatures is a real problem and changes in film structure are well-known. Internal reflections within the layers enhance the absorption of solar radiation by effectively increasing the thickness of the stack, but can also give rise to undesirable interference losses in the infra-red (figure 8.7). A layer of

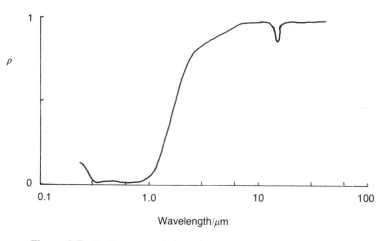

Figure 8.7. Reflectance of the selective absorber of figure 8.6.

silicon thinner than $2-5 \mu m$ would certainly show none of this infra-red anti-reflection and the emissivity would be low (as required), but then the solar absorption in the silicon would be less effective. More efficient absorbers (i.e. direct gap semiconductors) will have steeper absorption edges and can be used in thinner layers; double layers of semiconductor absorber may give greater efficiency, at the expense of additional complexity.

8.3. *Extraction of energy*

In order to use a flat-plate collector as an energy source there must be some thermal output from the heated absorber to a liquid circulated through it and a storage tank by a pump, or by natural convection (sometimes called a thermosyphon). (Some collectors have used a black fluid in transparent plastic tubing, to directly absorb solar energy in the fluid.) The actual path across the absorber may be in open channels, or in closed passages, for example a serpentine tube

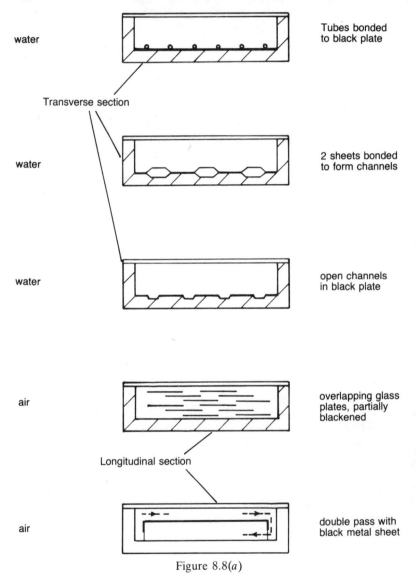

Figure 8.8(a)

bonded to the plate. The open-channel method may give condensation of vapour on the inside surface of the cover plate and requires the liquid flow to be maintained by a pump, rather than natural convection, but is the simplest method. Other designs must overcome the problem of producing an intimate thermal contact between absorber and fluid through the channel walls, by soldering, welding or adhesive. Corrosion occurs when mixed materials are used in a water system (e.g. aluminium and copper), through electrochemical action. The all-copper system has many advantages despite the initial cost, for it is a good thermal conductor, easily formed, easily bonded, and compatible with conventional plumbing. Some examples of flat-plate collectors are shown in figure 8.8 and in the photograph. Small bore tubing can have an appreciable pressure drop along even short runs, and if it is used for the feeder pipes to the collector a large circulating pump will be required to maintain a uniform flow across the whole width of the collector. Reverse circulation by natural convection must be prevented at night or heat will be removed from the storage tank.

There are many reports of comparative tests of flat-plate collectors, either in use for domestic heating or under solar simulator conditions. It has been found that collectors with two cover plates perform more efficiently than single-cover collectors at high temperatures, although

Figure 8.8(*a*). Cross-sections through some flat-plate solar collectors used for air or water heating. (*b*) Solar collectors undergoing comparative tests. (Reproduced by kind permission of Kent Solartraps Ltd.)

at lower temperatures a single-cover collector is better because of lower absorption in the cover. It is always important to transmit the maximum possible amount of solar radiation to the absorber. In general, an extraction efficiency of 50% is not uncommon for temperature rises of less than 40° C, but this is only a minimum temperature for domestic purposes, and higher fluid temperatures are achieved with some loss of efficiency.

Water storage tanks in directly-heated systems should ensure that the incoming hot water does not mix with the colder water which is about to be passed through the collector. Various ideas have been tested for protecting the plumbing against freezing of the water, including draining of the whole system whenever the temperature falls below 0° C, or the provision of expandable trunking. In an *indirect* circuit (figure 8.9) the collector fluid passes through a heat

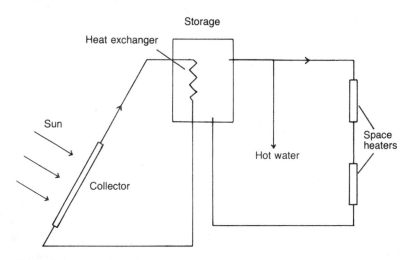

Figure 8.9. An indirectly heated storage tank and solar heater. Valving has been omitted from the figure. The system shown uses natural convective circulation.

exchanger in the hot-water cylinder and it is simpler to use a solution of anti-freeze. Perhaps this is the best answer, despite the additional efficiency loss due to the heat-exchanger and despite anti-freeze solution having a slightly lower specific heat capacity and thermal conductivity than pure water. Successful installations have used water heating and storage in the same unit to provide hot water for evening use. These simple and effective units do not require the same amount of technical expertise or materials as the flat-plate collectors discussed above. (Even a black plastic bag of water left in the sun all day will provide a useful quantity of warm water.) Of course, the design of a solar water-heater for domestic hot water or house heating depends greatly on the location and architecture, for attachment of solar heaters to existing houses is expensive.

The calculation of the outlet temperature of the fluid from a flat-plate collector in any given set of operating conditions is not simple. Manufacturers' figures will naturally err on the over-confident side but should be correct for their set of assumptions: it is quite common to quote performance for ideal AM1 insolation, rather than for a more realistic mean annual insolation. In an indirect cycle the heat-exchangers work more efficiently at *low* flow rates (as long as a stagnant layer of liquid does not form next to the walls and introduce an extra thermal resistance), whereas the collector itself 'prefers' *high* flow rates (to give a small increase in temperature and so keep down thermal losses). Somewhere between these extremes there is an optimum flow rate for transfer of heat to the storage tank. A thermosyphon is to some extent self-regulating, for low levels of insolation will automatically produce slow circulation of the fluid.

If we first assume that the fluid on exit is at a constant temperature equal to that of the absorber, it will have extracted energy at a rate P_{out}, and we can draw up a new set of thermal balance equations. The equilibrium temperature of the absorber will be lower than for the isolated case, which is equivalent to reducing the solar input. For an absorber plate separated from a glass cover by a vacuum space, we have the following thermal balance for unit area, when heat is extracted (see figure 8.10).

$$a_A(E + \varepsilon_g \sigma T_C^4) = \varepsilon_{A\sigma} T_A^4 + P_{out}. \tag{8.6}$$

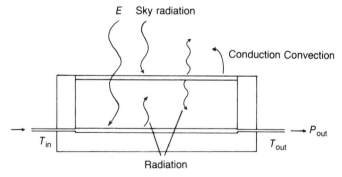

Figure 8.10. Heat extraction from an evacuated flat-plate solar collector, showing energy balance (W m^{-2}).

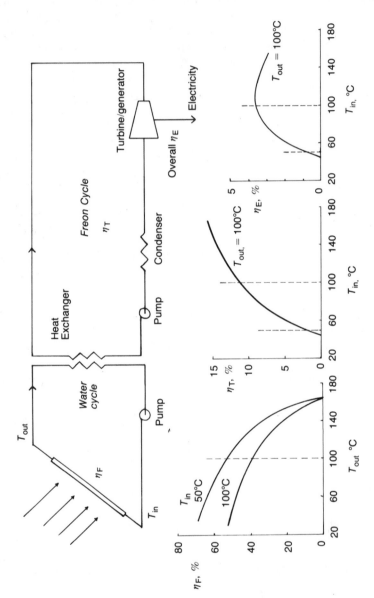

Figure 8.11. Efficiency of a flat-plate solar collector driving a heat-engine to produce electricity.

For the cover, equation (8.5), given previously for this collector type, remains unchanged. There are now three unknowns: T_C, T_A, P_{out}, for which there will be a certain equilibrium temperature, T_A, according to the extracted energy. This temperature is found to fall approximately linearly as P_{out} is increased. More significantly, the inverse trend shows that the delivery of high-temperature fluid results in low power-extraction (that is, a lower overall conversion efficiency) because the losses are greater. Thus we can write:

$$m \, c \, \Delta T = P_{out}, \tag{8.7}$$

where m is the mass flow, c is the fluid specific heat capacity, ΔT is the temperature increase of the fluid through the collector. Figure 8.11 shows the declining efficiency of heat extraction as attempts are made to deliver hotter liquid. Both indirect cycles and heat engines require high temperatures for high efficiency.

It is common heat-engineering practice to reduce the thermal balance equation of a flat-plate collector to a linear form, collecting all the *undesirable* energy losses into one term, the loss coefficient U_L, and using a 'heat-removal factor', F_R, for the *useful* energy loss. This is known as the 'Hotel–Whillier–Bliss' equation. The 'instantaneous rate of energy gain', Q, for unit area, is

$$Q = F_R[Ek\tau a - U_L(T_{in} - T_{out})], \tag{8.8}$$

where k is the shading and dirt factor, τ is the transmittance of the cover sheet, a is the absorptance of the absorber plate, T_{in} and T_{out} are the fluid input and output temperatures.

The combined efficiency of a flat-plate collector and heat engine is unavoidably low, as we can see by considering a system operating with a hot fluid at 473 K and rejecting heat to the atmosphere at 323 K. The efficiency of a Carnot engine operating between these limits is 32%, and a Rankine cycle engine would not give more than one third of this performance owing to the large temperature drop. If the *collector* operates with an optimistic 60% efficiency, the overall efficiency for collector and heat engine can be no more than about 6%. We have neglected to allow for heat exchanger losses, pump consumption, and control valving. A more realistic overall efficiency for electricity production would be 3–5% (see figure 8.11), if the turbine and generator were at least 90% efficient. At this efficiency, in the UK a total annual electrical output of $3\cdot6 \times 10^6$ GJ (that of a medium-sized power station) would require more than 1 km^2 of actual solar collector.

9. Concentrator collectors

A higher overall efficiency demands a higher working-fluid temperature than can be given by the simple flat-plate collectors described so far. Table 8.1 showed that only selective absorbers were able to give temperatures approaching the 500° C used in most steam-powered cycles, and yet these are *stagnation* temperatures with no heat extracted. Reducing the energy loss still leaves us below such high temperatures, and the only alternative is to increase the input power density by optical concentration.

Let us look at the stagnation temperatures which could be reached by an absorber with various degrees of concentration for a solar flux of 800 W m^{-2} (Table 8.1). The flat-plate collector with a vacuum interspace, when combined with four inclined plate mirrors to give a concentration ratio (CR) of 2, would now reach an equilibrium temperature of 478 K, or 829 K, according to the selectivity of the surface. The energy-balance equations are the same as before, with the input power simply multiplied by the CR. A *low* CR in our most complex flat-plate collector, with a black-body absorber, is not as effective as using a selective absorber in even a simple structure, which shows the great advantage of a selective absorber. It is obvious that concentrators will only be used where they give a sufficiently high value of CR to offset the additional cost and complexity. Winston collectors are a compromise here.

With a CR of 10^3, which is perhaps the most to be expected of real surfaces, equilibrium temperatures of 2302 and 3987 K are given by our simple energy balance equations. These are over-estimated because at such high temperatures we should investigate the thermal losses more closely; conduction along the struts supporting the absorber would be appreciable, and the pattern of collector in which a CR of 10^3 is reached would no longer look like our flat-plate arrangement. Such values *are* achieved by solar furnaces. Having seen the order of temperature to be expected from highly concentrating optics in direct sunlight (figure 9.1), we can now examine some of

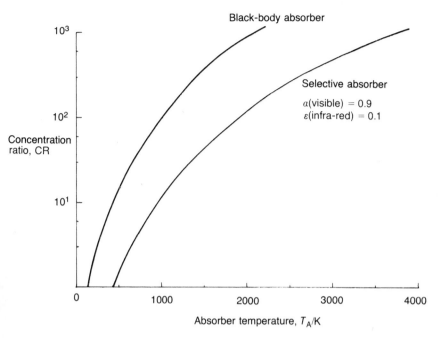

Figure 9.1. Expected stagnation temperatures of evacuated solar collectors with concentrators.

the actual structures proposed. One of the problems is that any cover window will itself reach high temperatures and would need liquid cooling. This might provide a secondary source of heat, useful in itself, or for preheating the main fluid. The absorption in any transparent window must be taken into account, since this can reduce the power incident on the absorber just as much as imperfect optical surfaces.

The work of early pioneers of practical solar engines (Mouchot, Ericsson, Shuman, Abbot, Eneas, and many others) is now being re-examined for its relevance to today's energy problems. Whilst parabolic trough collectors with a tubular absorber, after the style of early collectors, are being constructed by many modern groups, they can now use an organic transfer fluid, different from the working fluid in the engine, instead of heating water directly from steam. But it is in the availability of better engineering materials that modern solar energy conversion has the advantage over the pioneers efforts. Newer designs of trough reflectors tend to use 'envelope' optics instead of large reflecting sheets, in the same way that early large-aperture mirrors were made from small plate mirrors. A series of mirror strips are mounted so as to produce the envelope of a curved surface (a Fresnel mirror).

A further change from convention is to track the absorber and keep the concentrating mirror stationary: a design reported recently has been used to provide domestic heating in a test house in the USA.

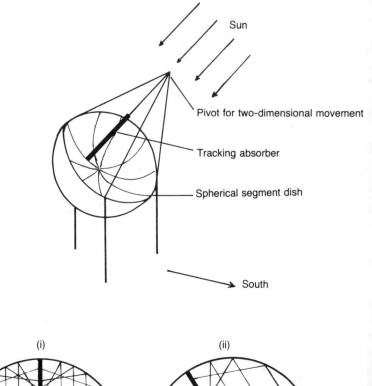

Figure 9.2. Stationary reflector/tracking absorber (SRTA) solar collector (described by W. G. Steward and F. Kreith).
Sections in plane containing absorber show ray paths, (i) at noon, (ii) two hours from noon.

Like all concentrator systems it is only effective in a climate having a large amount of direct insolation, unlike the UK. The arrangement is shown in figure 9.2. The reflector is a segment of a sphere, forming an image on a linear absorber which tracks this solar image across the interior of the dish. Details of the absorber diameter and length can be determined by geometry, according to the rim angle of the reflector. The absorber axis is always aligned with the solar rays passing through the centre of curvature, and so requires tracking in two dimensions. An advantage with this arrangement is that the reflector can be rigidly mounted to ensure it will withstand the worst that winds can do, whereas tracking dishes require a heavy tracking mechanism if they are to survive high winds. The absorber itself will not receive a uniform energy flux along its length, and this can lead to large thermal-expansion stresses.

An equally clever absorber-tracking collector uses a cylindrical mirror formed of long thin reflecting strips, each arranged on the circumference of a circle (when viewed from one end of the cylinder) in such a way as to form a solar image on the same circle circumference. The absorber is pivoted about the circle centre and traces the circumference as it follows the Sun's movement (figure 9.3). (This Russell collector uses the same principle as the famous Rowland diffraction-grating circle.) Seasonal adjustment is needed to avoid adjacent strips from shadowing others, and the acceptance angle is low.

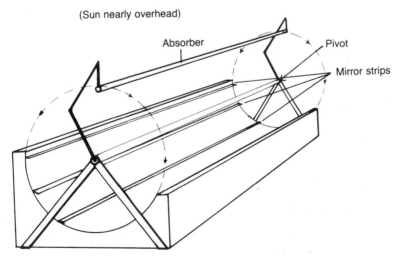

Figure 9.3. Tracking absorber solar collector designed by J. L. Russell. Only a few of the Fresnel mirror elements are shown.

The 'power tower' projects under way in the USA and in Italy synthesize a large aperture concave mirror from a number of small mirrors arranged in a two-dimensional array (figure 9.4). The absorber, mounted on the top of a tower positioned to the south of the centre of the mirror field, must intercept the reflector rays from

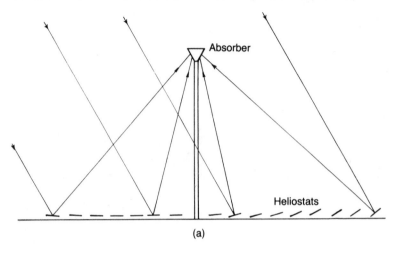

(a)

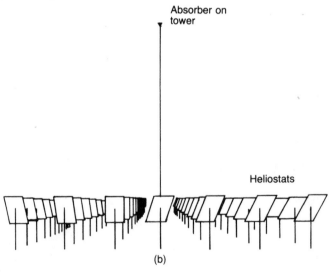

(b)

Figure 9.4. A 'power tower' solar collector. (a) The Fresnel mirror effect, and (b) the mirror field.

each mirror. The total solar input to the receiver is greater by a factor of about ten than the thermal flux in conventional steam boilers, and thus a cavity absorber with internal fluid passages is proposed to reduce radiative loss. A scaled-down version of one such boiler has recently been tested in France at the Odeillo solar furnace, as part of the development of a 100 MW (electrical) plant in the USA. The final power plant, to be sited in Arizona, will require over 10 000 36 m^2 heliostats. The receiver is to absorb 70% of the total solar energy incident on the mirrors, giving high-pressure steam at more than 5000° C. Since a large part of the total cost is the tower itself, the economics of one large tower versus several smaller towers should be at least as important as the problems of the geometrical optics. As one moves away from the tower, the angle of the heliostats to the horizontal increases, and so must their separation to avoid shadowing and blocking. Eventually the cost of adding further mirrors to the perimeter offsets the small increases in solar energy collection. A practical limit of 3–5 times the tower height is set on the field diameter. Whilst concave heliostats would increase the theoretical concentration ratio of the collector by reducing the size of the solar image, the increased production cost would be too great. The tracking of such a large number of mirrors could be partially by sun-sensor control, and partially by timed motor drive. Computer control would provide even lower tracking errors, and would allow the image at the collector to be 'shaped' to give different flux profiles across the surface.

Most concentrator collectors, and especially the power tower scheme, are inappropriate for Northern Europe, although the UK is involved in a project in Southern Europe through its joint solar energy programme. In the UK there would be collection of less than 40% of the annual insolation on the area if such a power plant was constructed.

10. Ponds and oceans

10.1. *Solar ponds*

High concentration ratio collectors are not suitable for large-scale use in the UK, as they waste so much of the diffuse radiation. One way to make fuller use of this is to cover a large area with our collecting medium. A technique for local on-site collection is the shallow roof pond. This collects solar radiation during the day by absorption in water, at the same time reducing the indoor temperature to tolerable levels in hot weather without air-conditioning plants. The heat may be retained in the pond during the night by an adjustable transparent cover, making the most favourable heat exchange path that with the building beneath. The feasibility of this design for domestic heating has been demonstrated in the USA.

An interesting natural heat collector and store exists in a few isolated lakes in Hungary. In these there is an inverted temperature gradient with heat trapped well below the surface, instead of the usual falling temperature with depth. The normal convective mixing process is opposed in these ponds by a density gradient produced by a dissolved salt concentration gradient. Although the term 'solar pond' includes all shallow bodies of water used to collect solar radiation, it is more commonly applied only to these non-convecting types. Let us examine the physical processes which allow bottom temperatures in these ponds to be maintained at 60–70 K above the surface temperatures (20–30° C).

In a normal shallow pond of uniform density, solar radiation is absorbed by the water and also at the bottom which is often blackened in energy-collecting ponds. (Infra-red radiation is absorbed within a few centimetres of the surface, but visible radiation penetrates to the bottom.) The heated liquid (especially away from the sides, which lose heat to the ground) will be of lower density and so will be displaced upwards by denser cold water. The effect of this convective mixing is that temperatures only about 2 K higher than ambient are achieved at the bottom of a shallow pond. The heated upper water layers lose heat by conduction/convection, and by evaporation to the atmosphere.

In a non-convecting pond, there is a concentration of dissolved salt which increases towards the bottom, giving a density gradient. Even when it is heated by solar radiation, the density remains greater than that of the colder liquid above. Heated liquid at the bottom of the pond will lose heat mainly by conduction to the surrounding ground, and convection losses will be almost negligible. Only a little heat will be conducted through the static layers above, to be lost by evaporation, convection, and radiation. Figure 10.1 shows the layers in a solar pond, and the heat flow paths. The bottom layers of the pond may be heated to near the boiling point of the solution ($>100°$ C) if the insolation is intense and no heat is extracted.

It is possible for the lowest region to be heated so much that the concentration-induced density *is* overcome by expansion, in which case convection currents will start. This convecting zone may extend only a short way upwards (say 20 cm in a total depth of 100 cm) and can, in fact, help to increase the storage capacity of the pond without destroying the stability of the system. A similar mixing zone at the *top* of the pond is detrimental since it increases the surface temperature and hence the heat losses. If this upper mixing zone is established it can reduce the useful energy extracted from the pond by one third, so it must be prevented, if necessary by transparent covers, or by groynes to limit wind convection and waves.

When a solar pond is started it is filled with layers of successively weaker brine and exposed to sunlight. Over a period of more than a year (depending on the thermal properties of the ground) it will build up to an equilibrium condition in which the initial stepped concentration profile becomes smoothed out by diffusion, and the bottom liquid temperature approaches a value dependent on the depth of the pond. A ten-layer pond 1 m deep takes about 100 days to reach an approximately stable linear concentration gradient.

Under ideal operating conditions, after the settling period, all of the solar energy absorbed within the depths of the pond should be extracted for useful work, whilst the body of the pond is warmed only by energy absorbed there. Suspended particles in the solution will tend to remain at a depth corresponding to their density, and can obscure the clear liquid to an extent that sunlight can no longer reach the bottom where energy is to be extracted.

Heat is most readily extracted by withdrawing the whole bottom layer of hot, dense solution at one end of the pond, passing it through a heat-exchanger, and returning it to the other end. Laminar flow is

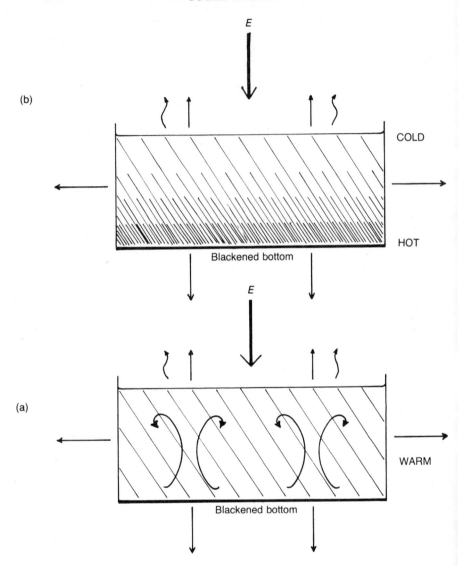

Figure 10.1. Solar ponds, showing energy loss paths: (*a*) normal pond with convection currents; (*b*) non-convecting salt pond. (Liquid density is indicated by the shading.)

possible in most large bodies of water, but is easier in the density-gradient pond, where mixing is deterred. Annual average efficiencies for the heat extracted, referred to the total solar input, approach 25% (in theory).

The stability of the inverted temperature gradient of a non-converting pond depends on the prevention of the upward diffusion of the salt. To overcome this drive towards uniformity, one could remove salt from the top of the pond and add it to the bottom, or else supply fresh water to the top whilst adding salt to the bottom (as in the natural ponds in Hungary). The necessary water addition is less than 1 mm/day. On the other hand, the extraction of energy by withdrawing liquid and returning it in a continuous cycle gives a situation where greater adjustment is possible, and this is essential because we neglected to allow for the moisture *lost* by evaporation from the surface (which can be up to 10 mm/day) and water *added* by atmospheric precipitation. The economic feasibility of solar ponds depends mainly on the cost of the salt (which need not necessarily be sodium chloride), both for initial installation and for maintenance.

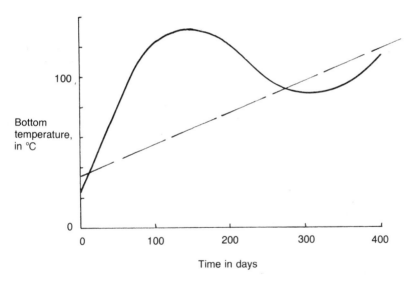

Figure 10.2. Building-up of 1 m deep solar-pond temperature after filling in spring with layers of successively more dilute brine.

The initial variation of temperature with time for a solar pond filled in spring is approximately sinusoidal, superimposed on a slow linear increase (figure 10.2). The large heat capacity of the pond should allow no short-term temperature fluctuations. This can affect the microclimate close to the pond, for the greatest energy loss will be in autumn when the air temperature is falling, and the pond yet retains summer heat. An example of the sort of difficulty this might create was found by H. Z. Tabor when an early solar pond in Israel developed gas bubbles due to the increased bacterial action in the bottom of the heated pond.

The temperature rise at a given level within the pond does not depend on the depth of the pond, but the bottom temperature obviously does. To some extent the higher steady-state temperatures are reached in the deeper ponds up to 2 m, but the radiation collection efficiency falls with increase in depth. Ponds 1 m deep are a workable compromise. Notice that these moderate temperatures allow only a low thermodynamic limiting efficiency for the conversion of heat to work (less than 4% in a practical system), similar to that of flat-plate collectors.

10.2. Ocean thermal energy converters (OTEC)

A large volume of solar collector already exists as the Earth's oceans. These have a temperature gradient in the expected direction, and in the tropics or at the poles this is sufficiently large to drive a heat engine, with low efficiency. The capital cost, operational lifetime, and environmental aspects may not be as attractive as the 'fuel' costs. With a 22 K temperature difference between surface and depths, such as exists in warmer ocean areas than the North Sea, the Carnot efficiency is around 7%. This is obviously very low, and comparable to that expected from a flat-plate collector. In fact, by the time the overall efficiency has been reduced by using a practical engine (operating on a Rankine cycle, say) together with heat exchangers, the proposition might seem hopeless! One major difference between these two heat sources is that solar energy arrives with a low power density, and requires a large acreage of flat-plate collectors, whereas an ocean thermal gradient source can operate with a small area collector by pumping sufficient warm water through the heat collectors. Indeed, the attraction of the solar sea-power plant lies in its present-day engineering feasibility and possible competitive cost with fossil-fuel power stations.

Not only has this scheme been discussed spasmodically for nearly a century, but a working plant was constructed by G. Claude (known for his work on the liquefaction of gases) in the Caribbean in 1930. This plant (figure 10.3) was less successful than his small-scale experiments which preceded its construction, partly due to heat losses in the insulated pipe which was to bring up cold water from more than 1500 m below the surface, so as to obtain a temperature difference of 28 K. By using surface water in the boiler and deep-sea water in the condenser, with rejection of the used water at an intermediate level

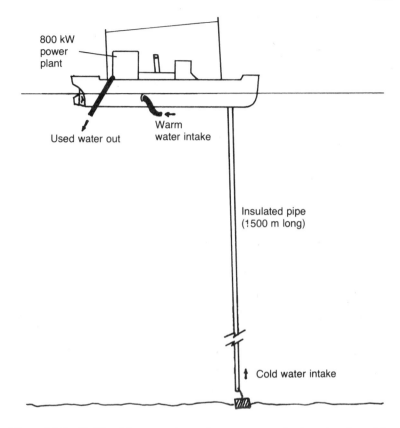

Figure 10.3. G. Claude's ocean thermal energy converter based on board *La Tunisie*.

into a convenient current, it had been hoped to generate at least 500 kW of electrical power. In practice, the pipeline was much shorter (only a 15 K difference was achieved), and the pumps and auxiliary equipment consumed more power than was generated. The pump capacity could have been reduced by using in the heat engine a working fluid, such as sulphur dioxide or ammonia, which has a much higher vapour pressure at these low temperatures than sea water (which would then be merely the heat-transfer medium). This had, in fact, been the plan of J. A. d'Arsonval 50 years previously: to use warm spring water to boil a fluid and cold river water to condense it, or even to use ice (a glacier) as the sink and running water as the source. The experiment by G. Claude and P. Boucherot used only the low vapour pressure of water itself in an open Rankine cycle, boiling in a partial vacuum. A further problem with sea water is its corrosive nature, and the dissolved air which is released on reaching the low pressure boiler. The two men tried a final large-scale experiment near Brazil in 1934, using a ship to house the surface plant, hoping to demonstrate success by operating an ice-plant from the generated electricity, but similar limitations kept the efficiency low and the ship was sunk in disgust!

A scheme with perhaps more chance of success, avoiding the long insulated pipe and pumps, was that suggested by Barjeot in the twenties. He wanted to use the Arctic seas for the boiler and an air-cooled condenser, with butane as the working fluid. Butane and sea water would be mixed in the boiler, and the resulting vapour would be condensed by a salt/ice mixture after passing through a turbine. Claude financed his own experiments, but no backer was forthcoming for this alternative plan.

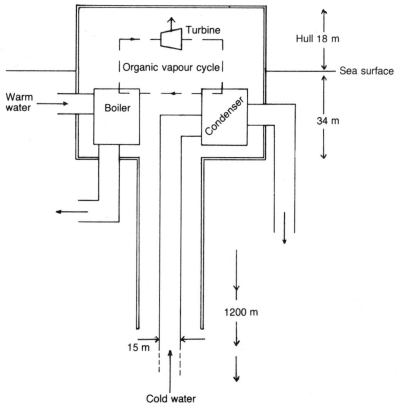

Figure 10.4. (a) Schematic design of an ocean thermal energy converter.

Most of the present schemes under investigation in the USA are similar to Claude's floating island concept, but use ammonia, propane, or a Freon in a closed Rankine cycle to drive a turbine. The transport of power from an island to the mainland presents some difficulties, but it has been suggested that hydrogen produced by electrolysis of sea water might be a better energy transporter than electricity. It is also possible to produce chemicals on site with the electricity.

The proposals present similar ideas, but differ in their choice of construction materials and techniques for maintaining a fixed position. The relative position of boilers and condensers depends on the

Figure 10.4. (*b*) Model of proposed ocean thermal energy converter. (Reproduced by kind permission of TRW Inc., California, USA.)

fluid and on the heat-exchanger design. A French-designed plant was tested off the Ivory Coast in 1956, and a large USA plant is to be constructed in the near future. The Anderson team have proposed a 100 MW generating plant, and have formed a private company in the USA to study such installations. Their designs place the boilers below the condensers to make use of hydrostatic pressure to contain the higher pressure vapour without heavy steel plate. Most other designs place the condensers lower, near the cold-water intake pipe. Some designs use a semisubmersible hull, whilst others use a surface hull (figure 10.4). The vital components in all designs are the pumps, which must continuously handle a large flow-rate of sea-water, and the heat-exchangers in which heat must pass very efficiently to the working fluid in the boiler to evaporate it, and from the vapour to the cooling water in the condensers. Present heat-exchangers are too expensive and would have to be made very large to offset their relatively low performance. The thermal impedance of a *smooth* surface with its associated fluid film is high, although the impedance to fluid flow past it *is* desirably small. In order to improve the transfer of heat across the walls, ridges may be introduced to break up the fluid film, but this will increase the flow resistance, and so require larger pumps. Corrosion is a problem, and marine fouling has to be prevented.

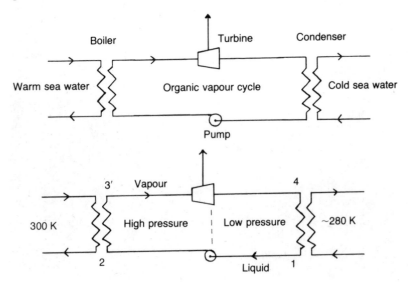

Figure 10.5. Closed Rankine cycle proposed for OTEC (compare with figure A.1).

We can estimate the warm-water flow-rate for a 100 MW generating plant, using the Anderson's propane cycle or the Zener ammonia cycle (figure 10.5). The energy balance is:

$$P_{out} = P_{in} \times \eta_R, \tag{10.1}$$

where $P_{out} = 100$ MW; $P_{in} = mc\,\Delta T$, with density of sea water, $m = 1025$ kg m^{-3}; specific heat capacity of sea water, $c = 4200$ J kg^{-1} K^{-1}; temperature drop through boiler, $\Delta T = 23$ K; and $\eta_R =$ Rankine cycle efficiency.

If Q is the flow-rate in m^3 s^{-1}, we find on substitution of the numerical values:

$$Q = 11 \cdot 7 / \eta_R \text{ m}^3 \text{ s}^{-1} \tag{10.2}$$

This indicates the importance of using a high efficiency working fluid cycle. This rate far exceeds that used for cooling water in conventional power plants, but the pump requirements may be reduced by siting the plant in a strong ocean current, like the Gulf Stream. (The kinetic energy of the exhaust water can be used to help maintain the hull position.)

The final question to be answered before this type of plant can be put into operation is not an engineering one, but is the important question of the environmental impact made by the change in water temperature and in ocean currents. A particular problem is the discharge of cold water close to the surface, producing a net reduction in the upper layer temperature, and perhaps with drastic effects on animal and vegetable life in the sea. This exhaust water will result in a *lower* energy loss locally from the sea to the air, especially by vaporization, and hence in a lower air temperature. As a result there might be a change in the convection current pattern, with a possible warming of the sea away from the power plant site, as the solar energy must be dissipated elsewhere. The actual changes expected must be closely modelled before installation commences.

11. Solar cells

11.1. *Sources of inefficiency and their improvement*

We have seen how ideal photovoltaic cells would be expected to behave, and some unavoidable sources of inefficiency have been discussed. We have yet to examine those factors which are within the control of the producer. These tend to interact and thus independent optimization of each is not going to result in the best possible cell. For efficient photovoltaic conversion, it is necessary to offset a loss in efficiency at one stage in the process by a gain in another stage, and to see how this trade-off arises we must examine optical absorption and creation of carriers in more detail. Then it will be possible to understand the reasons for the variations on the theme of a basic p–n junction cell, before examining some alternative, potentially cheaper, junctions.

Listed on p. 126 are the sources of conversion inefficiency for a silicon cell in AM1 insolation (and see figure 11.1). An upper limit of about 22% has been suggested for the power conversion efficiency of these cells (AM1 illumination), for 23% of AM1 illumination consists of photons with energy less than the bandgap, and a further 33% is lost because of incomplete utilization of the eligible photons above the bandgap energy (figure 11.2). The 44% remaining, subjected to a voltage factor (VF) of 0·5 gives 22% efficiency.

11.1.1. *Optical absorption*

If it were possible to utilize the portion of the solar spectrum with photon energy below E_g, then the overall efficiency could be increased whilst still using silicon. Two schemes have been suggested, but in practice each provides new loss paths, and it would seem impossible to extend the long wavelength response of a cell without changing the semiconductor. These theoretical improvement schemes are as follows:

(*a*) The addition of an 'up-converter' phosphor to the top surface of the cell, to absorb two low-energy photons and produce a single

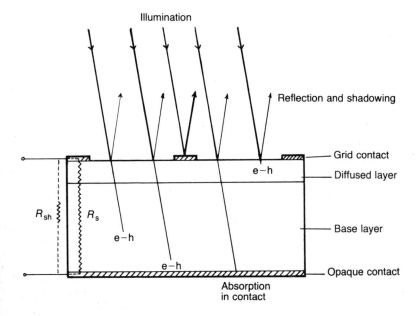

Figure 11.1. Losses in photovoltaic cells. e–h signifies production of free electron–hole pairs and subsequent recombination.

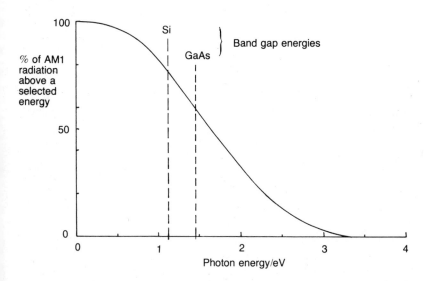

Figure 11.2. Percentage of AM1 insolation capable of producing electron-hole pairs in a chosen semiconductor.

Losses basic to the material chosen	Losses which are mainly process limited
(1) Loss of all photons with energy below E_g.	(1) Reflection of incident illumination at top surface.
(2) Quantum efficiency is less than 1.	(2) High front surface recombination. (This affects the short wavelength illumination.)
(3) Absorption coefficient is not infinite for all wavelengths. (Some photons with the correct energy are still not absorbed unless the material is very thick.)	(3) High rear-surface recombination. (This affects long wavelength illumination, especially in thin cells.)
(4) Each photon with energy above E_g produces only one electron-hole pair. (The excess energy appears as heat.)	(4) Bulk recombination due to poor minority carrier lifetimes.
(5) Voltage factor less than 1.	(5) Recombination of carriers at the junction, and tunnelling currents, which are shown by a diode factor, n, greater than unity.
(6) Curve factor less than 1.	(6) Series resistance of the bulk, the contacts, and the thin diffused top layer of p–n cells.
	(7) Shunt paths across the cell due to surface films and junction defects.

higher energy photon, for subsequent absorption in the semiconductor. Such two photon absorption processes are inefficient except at high intensity levels (e.g. laser powers), but have been used in infra-red image converters.

(*b*) The addition of impurities which have energy levels within the energy gap, and so can act as intermediate steps for the transfer of electrons from valence band to conduction band by the absorption of more than one photon. In fact these levels have an adverse effect on the device behaviour because they also act as intermediate steps for the *return* of electrons from the conduction band to the valence band: that is, they increase the recombination rate more than they increase the generation rate.

This loss remains, whatever bandgap is selected, and the best way of reducing it might be to use a stack of several different semiconductors. If these were graded from the illumination face to the back contact in order of decreasing energy gap, then low-energy photons would pass through the stack until they reached a layer with bandgap narrow enough for them to be absorbed. Apart from the difficulty of producing such a complicated sequence of materials, this seemingly elegant method has a drawback which shows up when we think about the electrical matching of one material to the next. Each cell must pass the same current if they are all in series, and yet each will have a different V_{oc}, I_{sc} and P_{max}. This suggests that the *overall* efficiency would be restricted.

The only successful approach to this concept is the 'graded gap' cell in which the energy gap decreases smoothly away from the top surface. This *can* be constructed: with a mixture of two semiconductors which can form a solid solution over a wide range of compositions such as GaAs and GaP, or GaAs and AlAs. The minimum bandgap of the latter pair changes from direct at $1 \cdot 43$ eV (100% GaAs) to indirect at $2 \cdot 2$ eV (100% AlAs) (figure 11.3). A solution of initially gallium, aluminium and arsenic could be used to produce a mixed composition crystal, with composition changing to pure GaAs by control of the growth conditions, or by altering the solution composition. A simpler way would be to start with a single crystal of GaAs and use this as a 'seed' to be coated with the varying composition $Ga_{1-x}Al_xAs$.

This property of the two semiconductors of being structurally compatible has an important consequence for GaAs cells, which are generally rather inefficient at high photon energies due to their high absorption coefficient and high surface-recombination rate. This means that many electron-hole pairs are created close to the surface and recombine rapidly before they can be separated by the junction field, unless the surface layer can be given a low recombination rate.

11.1.2. *Voltage factor*

An examination of the conditions needed to make VF as large as possible shows that for an n-on-p silicon cell the limit is $0 \cdot 76$. Imagine that the V_{oc} *is* equal to the bandgap. We saw earlier that the open circuit voltage of an ideal diode depends on the ratio of I_L to I_o (equation (7.17)). For any chosen illumination source the value of I_L is a constant, and we should have to make I_o equal to 2×10^{-17} A m^{-2} for $V_{oc} = E_g$, if I_L was as large as 500 A m^{-2}. If I_L was (realistically) less than this, I_o would have to be even smaller. Such a low value of I_o can only be reached by having a very high concentration of dopants in the silicon, together with very long minority carrier lifetimes (equation (3.9)). The value of the diffusion

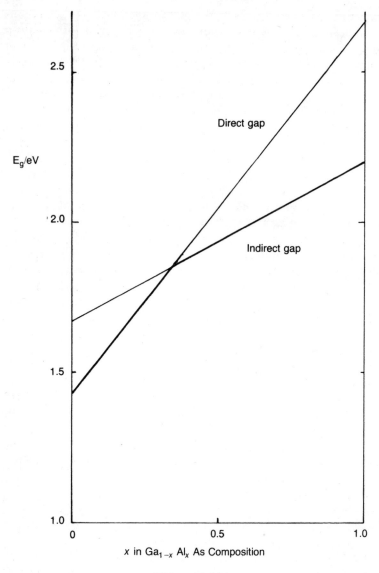

Figure 11.3(*a*).

coefficients in this equation depend directly on the carrier mobility, and this would unavoidably be low in silicon doped as highly as it would have to be for the value of I_o suggested. Even with high doping and low mobility, τ_p and τ_n would be reduced from a reasonably high value of 10^{-5} s (in lightly doped silicon). In addition, the other

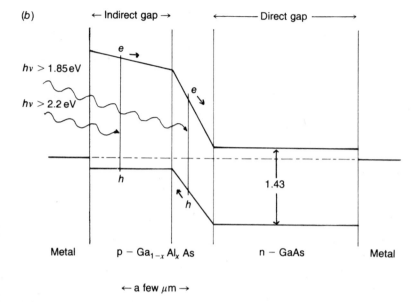

Figure 11.3(a). Variation of bandgap E_g with value of x in Ga_{1-x} Al_x As. (b) Simple model of energy levels across a graded gap cell based on GaAs. e are electrons, h are holes.

sources of leakage current, and especially recombination current at the junction, would be larger than the ideal value. So if the diode was made from highly doped silicon (to move the Fermi level to the conduction band in the n-type side, and to the valence band in the p-type side), the very high concentration of impurities would *increase* the leakage current, producing a saturation of V_{oc} rather than an increase. Indeed, if the doping were too great, the energy levels produced by the donors and acceptors would be sufficiently numerous to reduce the energy gap, which also would help to increase I_0. In practice, doping levels of 10^{21}–10^{22} m^{-3} are commonly used, which produce resistivities of $0\cdot1$–$0\cdot01$ Ω m and a VF of $0\cdot5$. (I_0 is then 10^{-8} A m^{-2}.)

11.1.3. Collection efficiency

Let us now turn to the third largest loss: this is the collection inefficiency of the photo-created electrons and holes. This term lumps all those losses of carriers by recombination, listed as numbers 2, 3, 4, 5 on the right-hand side of our 'debit' table. Some recombination of charge pairs takes place before they have reached the junction, but a large component of recombination current occurs at the junction in poor cells. The recombination of photocarriers obviously represents a loss of current to the external load. Notice that we have placed this in the 'process limited' column since, in theory, there could be no loss of carriers in a cell. In many devices (e.g. transistors) it is advantageous to have a rapid recombination of minority carriers, so that the device responds rapidly to changes in stimulus, and as a result there has not been a great effort in understanding the causes of bulk recombination other than attributing it to defects. Certainly it is true that the lifetime of silicon depends on the method used to grow it, and on subsequent treatment. The best silicon at present has lifetimes of more than 100 μs, which could give very good collection efficiencies (and V_{oc}) in a solar cell if the other parameters of the silicon were suitable for cells. The two surface recombination processes are generally defined by a recombination velocity, which is very high in GaAs (10^5 m s^{-1}) compared with silicon (10^3 m s^{-1}).

There are two ways of overcoming this limitation. The first, relevant to GaAs cells, is the addition of a thin layer of $Ga_{1-x}Al_xAs$ (with x greater than 0·35) so that the 'surface' of the GaAs is not then an abrupt one, and yet photons of energy above the bandgap of GaAs can pass through this $Ga_{1-x}Al_xAs$ 'window' to be absorbed close to the cell junction. The discontinuity in the conduction band (see figure 11.4) prevents electrons created in the GaAs from reaching the outer surface.

The second technique is to add a field within the cell away from the junction region to drift minority carriers away from the contacts. This is more often applied to the back contact, and is then known as a back surface field (BSF). It is produced by adding a thin heavily doped layer between base and metal contact, and has a secondary purpose in reducing contact resistance (figure 11.5). It may be used in association with drift fields in base and top diffused regions, both of these being produced by a non-uniform dopant concentration with depth. The small barrier, ΔE_c, at the p/p$^+$ junction prevents electrons from easily reaching the back contact, and so effectively reduces the rear-face recombination velocity (to 0·1 m s^{-1} from the infinite velocity of an ohmic contact). This is especially important in thin cells, in which carriers would be created in large numbers near the back contact.

The short-wavelength response can also be improved by having the junction as close to the top surface as possible, in order to separate carrier pairs before they have time to recombine at the surface. These violet cells (so-called from their spectral response) have junctions

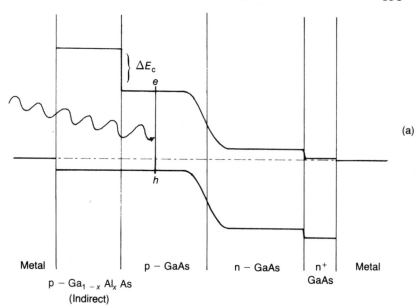

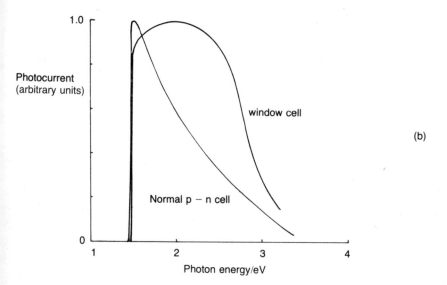

Figure 11.4. GaAs p–n cell with $Ga_{1-x}Al_xAs$ window. (a) Energy levels; (b) spectral response.

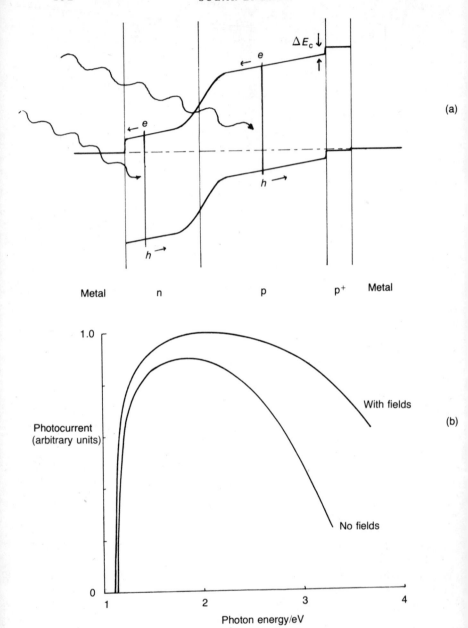

Figure 11.5. Back surface field and drift fields added to Si p–n cell. (a) Energy levels; (b) spectral response.

only 100–200 nm below the surface instead of 300–400 nm deep, and eliminate the surface 'dead layer' by using a lower surface concentration of dopant before it is diffused into the crystal. The electrical resistance of this thin diffused layer is higher than for conventional cells and must be compensated by a finer top-grid pattern.

The way to improve the collection efficiency is, then, to use a shallow junction, with drift fields in the base, and with a large minority carrier lifetime throughout: not an easy cell to fabricate. Present cells have collection efficiencies of 70%, and this might be increased to over 80% by implementing these improved structures and increasing τ.

11.1.4. *Curve factor*

We now turn to the losses grouped under the term 'curve factor', which is already approaching the ultimate to be expected (88%). The CF has been introduced earlier as a measure of the squareness of the *I–V* characteristic, when series and shunt resistance losses are eliminated. It arises because the diode characteristic has an exponential shape rather than the step required for a CF of unity. This means that it is impossible to achieve an operating point given by V_{oc}, I_{sc}. The reduction is due to the forward current of the cell, I_D, which in turn depends on the ideality factor, n, and the leakage current, I_0. As a result, CF increases with V_{oc} in ideal cells, and so can be seen to increase with E_g. Real cells have leakage currents arising from the other paths discussed, as well as from diffusion, and so have a CF less than that expected from the perfect cell with $n = 1$.

11.1.5. *Reflection loss*

Finally, we have the process-dependent variables remaining on the right-hand side of the table. The reduction of reflection losses at the entrance surface of the cell is effected by adding at least a single layer of a dielectric with selected refractive index, n_D. Without such a layer the losses amount to as much as 34% at 1100 nm, rising to 54% at 400 nm, in silicon. A single-layer coating can reduce the reflection loss to 6–9%, and a double-layer coating can reduce the loss to 3%. To reduce the reflectance at a chosen wavelength to zero, the refractive index of the coating must equal the square root of the refractive index of silicon (which is 3·9 at 600 nm). Since we have broadband illumination for solar cells, and since n varies with wavelength, it is impossible to prevent reflection losses for the whole solar spectrum, but by using multilayer coatings it is possible to broaden the low-reflectance band (figure 11.6). The cost and com-

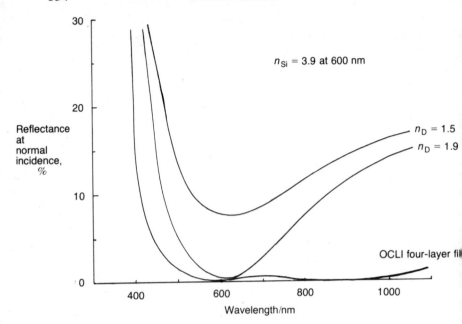

Figure 11.6. Anti-reflection coatings on silicon (n_D is the refractive index of the coating). (Based on data from OCLI, California, USA.)

plexity of these multilayer coatings must be weighed against the incremental improvement in cell efficiency, as compared to that given by a single quarter-wave coating.

A more easily understood way of reducing reflection losses is to roughen the top surface of the cell in a controlled manner, although this does not produce the low reflectances characteristic of dielectric layers. The technique avoids the use of the expensive equipment needed to deposit thin dielectric films. Instead a chemical etch (e.g. NaOH or hydrazine hydrate) produces small pyramidal facets on the silicon surface, each facet inclined so as to reflect light on to an adjacent facet if it is lost from the first attempt at penetration (figure 11.7). The junction itself follows these surface contours to make the most of the enhanced carrier production. Despite the difficulties of maintaining a low surface recombination velocity, and obtaining a good top contact, cells with AM0 efficiencies of 15% have been reported. This is an improvement of 4–5% over the performance of production-type n-on-p cells.

A similar roughening treatment on the back surface is highly undesirable, although it is often part of the process of forming a good

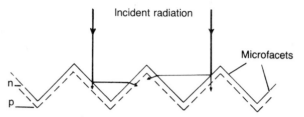

Figure 11.7. 'Comsat' non-reflecting (CNR) cell with textured surface and shallow junction, showing reduction of reflection loss.

contact. If the back surface of the silicon was highly reflecting, especially to long wavelengths, then light which would normally be wastefully absorbed in the metal contact could be reflected back into the cell for further useful absorption closer to the junction. This would be especially significant in thin cells, which are manufactured from low lifetime material. A suggested type of surface texturing for both faces is the production of 'V' grooves to reduce front surface reflections, or to reflect transmitted light from the rear surface back into the cell.

11.1.6. Resistance losses

There now remain the two resistance losses common to all cells. The series resistance arises from separate parts of the device and is distributed throughout the cell, although it is quite common to use a single lumped resistance in cell analysis. The two main causes of series resistance in modern cells are the thin top diffused region, and the top metal contact. There is an additional small contribution from the base region. Now is the place to examine in more detail the physical structure of a solar cell (figure 11.8).

The n-on-p cell is made from slices of single-crystal silicon doped with acceptors (e.g. boron), and with a thin top region doped by diffusing in donors (e.g. phosphorus) at a higher concentration than the already-present acceptors. This diffused region must be thin for good short-wavelength response, and so it will have a large effective resistance for carriers (electrons) moving parallel to the surface, in the plane of the layer. If this region is heavily doped to make it n^+, it will have a short lifetime for the minority carriers, and the depletion region will be narrow. Contacts are formed on the top and bottom faces of the cell after a careful chemical etch, and are alloyed with the silicon by thermal annealing. The base contact can cover the whole back surface and be opaque, so it is thick enough to present no resistance, but the upper contact must allow light to enter the cell. In

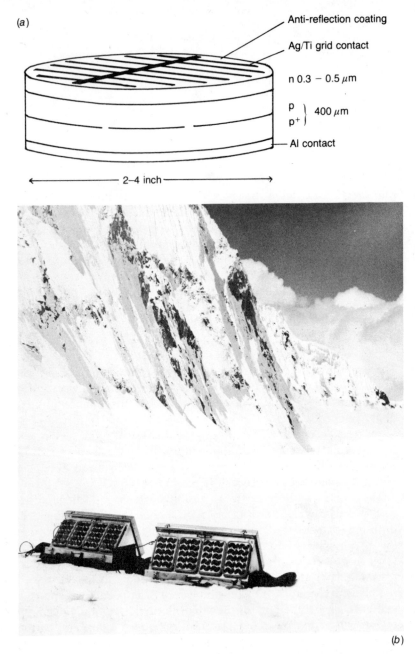

(a)

Anti-reflection coating

Ag/Ti grid contact

n 0.3 – 0.5 μm

$\left.\begin{array}{l} \text{p} \\ \text{p}^+ \end{array}\right\}$ 400 μm

Al contact

2–4 inch

(b)

Figure 11.8(a). Terrestrial pattern silicon n–p solar cell. (b) Ferranti solar-cell array used to recharge camera batteries on Everest. (Reproduced by kind permission of Ferranti Ltd, Manchester, UK.)

Schottky cells, this top metal layer forms the built-in field and so covers the whole face, but combining semitransparency with low electrical resistance requires careful design. In p–n cells it is more usual to use a gridded top contact (and one of these might be added to Schottky cells as well). The violet cells have a very thin diffused region and require as many as 60 closely spaced grid stripes for a 2 cm × 2 cm cell to reduce its series resistance below 0·1 Ω. In general, the contact covers 8% of the cell surface. The early space cells often had a series resistance of 0·25 Ω for a 2 cm × 2 cm cell, which meant that if the current drawn was 300 A m^{-2}, the voltage drop across the contacts was 30 mV, which is not an inconsiderable loss. In normal solar-radiation intensities the series resistance is dominant, although at low light intensities the shunt resistance may affect the cell voltage.

If this cell was used in a concentrator arrangement, giving a greater current output, the lost voltage would be proportionally greater. Also important at high intensities is the internal Joule heating. This, together with the dissipated part of the energy absorbed from photons of energy greater than E_g, will at very high concentration ratios reduce the cell efficiency by increasing I_o. (If the cell temperature rises too far there may even be permanent contact degradation.) For silicon p–n cells the measure of the efficiency degradation with increasing temperature is $-0·05\%$ K^{-1}, mostly through a drop in output voltage.

11.1.7. *Present performance*

Now let us examine the state-of-the-art of cell efficiencies. Table 11.1 gives the AM0 and AM1 efficiencies for some well-advanced structures, but is certainly not exhaustive. Some promising *new* techniques will be compared in a later table. n-on-p silicon cells are preferred to p-on-n because the transport of minority carriers is easier in p-type silicon. This means that the all-important diffusion of minority carriers in the base region is more efficient for electrons in p-type silicon than for holes in n-type silicon. The top diffused region is generally thin enough for both carrier types to move with comparable ease, or difficulty. Another consequence of the doping chosen for the base region is the dependence of the open circuit voltage on base resistivity. Notice that in the table the value of V_{oc} increases for the three values listed for n-on-p cells, but that this has not produced the expected monotonic increase in efficiency, since the minority carrier lifetime is reduced at high doping levels, and leakage current is increased.

TABLE 11.1. *Solar cell efficiencies (%) (best: η and theoretical: η').*

Cell	AM0		AM1		
	η	η'	η	η'	V_{oc}(mV)
n/p Si (0·1 Ω m base)	11·5	18	14	20	550
n/p Si (0·01 Ω m base)	13·0	20		21	600
n/p Si (0·001 Ω m base)	9·0	21		22	700
n/p Si (0·1 Ω m base and BSF)	1% improvement				
n/p Si (0·01 Ω m base) violet cells	14·0		17		
n/p Si (0·01 Ω m base) CNR-a textured surface cell	15·5		19		
p/n Si (Li-doped)	13·0		15		
p/n GaAs	10·0	23	12	25	900–950
p-Ga$_{1-x}$Al$_x$As/p-GaAs/ n-GaAs	12·5		16		950
p-Cu$_x$S/n-CdS	8·0		10		500

Since the absorption of light in silicon is low, it might seem desirable to use a fairly thick cell to ensure maximum solar absorption. A very thick cell would certainly have the rear ohmic contact well separated from the maximum depth of penetration of even the longer wavelength photons, but would also have increased series resistance. More importantly, the finite lifetime and hence finite diffusion length, L_B, of the minority carriers in the base makes it senseless to attempt the collection of carriers created more than about $2 L_B$ from the junction. In $10^{-3} \Omega$ m silicon the lifetime is 10^{-6} s and L_B is 50–100 μm, thus an increase of silicon thickness beyond 200 μm is largely wasted. At the other extreme, a thin silicon cell only 1–2 μm thick will not have a great absorption efficiency unless multiple internal reflections are arranged (figure 11.9). Direct-gap materials, on the other hand, can be made into thinner cells without penalty.

11.2. *New cells*

Now we shall move away from conventional silicon technology to examine the merits and defects of:

(a) Other semiconductors with better optical absorption.

(b) Cheaper fabrication methods using thin semiconductors.

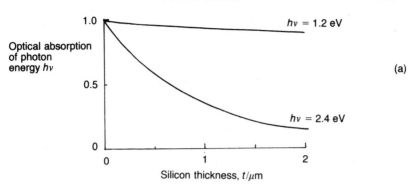

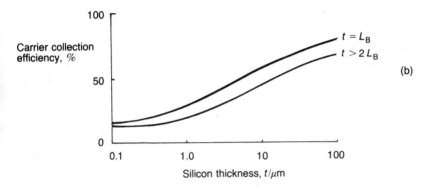

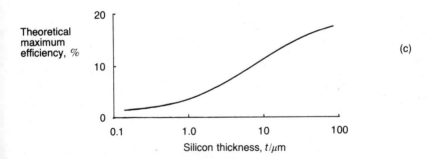

Figure 11.9. The significance of solar cell thickness. (L_B is the diffusion length in the base.)

(c) New structures such as vertical multi-junctions, grating cells, and MIS cells.

11.2.1. *Concentrator cells*

For terrestrial surface installations the main contest is between a large array of cheap, but only moderate efficiency cells, and a smaller number of expensive high efficiency cells, operated under large concentrating-optics arrays and possibly requiring solar tracking. Figure 11.10 illustrates concentrator solar-cell arrays combined with thermal collection of solar radiation through a cooling fluid. GaAs cells are preferred for high concentration-ratio collectors to avoid the need for much cooling, because they stand up better to high temperatures. For only moderate concentration ratios, silicon cells with a low series resistance and special grid contact are acceptable.

One of the chief alternatives to silicon p–n junctions is, therefore, based on gallium arsenide, with its direct gap of 1·5 eV (close to the optimum value) and high absorption coefficient. A problem with GaAs is the very high surface recombination velocity since GaAs is not easily coated with a passivating oxide. An effective way of dealing with this is to overlay the p–n cell with a thin 'window' of $Ga_{1-x}Al_xAs$ $(x > 0·35)$ to give a top surface which has an indirect gap (figure 11.4). The resulting cell is in general a p–p–n cell due to the diffusion of acceptors from the $Ga_{1-x}Al_xAs$ into the GaAs. Cells of up to 17% AM1 efficiency have been produced, and are attractive for concentrator use since their efficiency can increase to over 20% at high CR levels. The series resistance is kept below 0·03 Ω for 1 cm^2 cells by using a very thin layer of the alloy material (which is not highly conducting). The concentrator helps to offset the high cost of producing this semiconductor, but limits the application to areas where direct sunlight is readily and often available. .

High surface recombination is also deleterious in the vertical multi-junction (VMJ) cell, where the illumination is parallel to the junction plane (figure 11.11). These cells are not easy to analyse because the junction is not illuminated uniformly. This must lead to a potential gradient in the vertical direction for each cell module, and so an undesirable internal current will flow unless the wafer is made very thin. Reducing the wafer thickness also reduces the junction area, and the corresponding dark current. (The open circuit voltage, however, will be controlled by the leakage current at the *worst* point). Even long wavelength radiation, just above the bandgap in energy, will make a contribution to the photocurrent in VMJ cells because although it is absorbed well away from the top surface, the junction is

(a)

(b)

Figure 11.10 (*caption overleaf*)

(c)

Figure 11.10(a) and (b). 1-kW array of silicon cells using Fresnel lenses with a CR of 50. This system simultaneously delivers 5-kW thermal power. (Reproduced by kind permission of Sandia Laboratories, New Mexico, USA.) (c) Array of GaAs solar cells using concave mirrors with CRs of 10^3. The cells operate at $100°C$, and deliver 1 kW. (Reproduced by kind permission of Varian Associates, California, USA.)

now close to the site of carrier creation. The elimination of a thin diffused layer parallel to the top surface should reduce the series resistance and so allow operation at high intensities. If heavy doping is used, the series resistance of the bulk will be a limiting factor, for it *is* now possible to use heavily doped, low-lifetime material as long as the distance between junctions is small.

The interesting feature of a series-connected array of units in a VMJ cell (figure 11.12(b)), is the high V_{oc}. A 100-junction cell would be expected to give around 50 V. Smaller numbers of junctions have been constructed, and operated with concentrators to give more than 6% efficiency. Theoretical calculations predict a limiting efficiency of 20% for a 500-junction device in $10^{-3}\,\Omega$ m silicon. The obvious problem with the VMJ is the costly and elaborate fabrication process

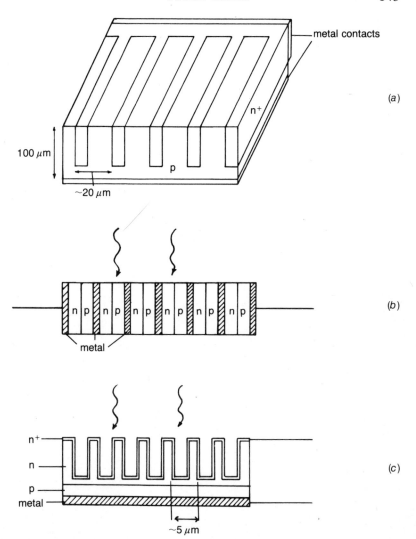

Figure 11.11. Vertical multi-junction (VMJ) cells. (a) Equivalent to several cells in parallel; (b) several cells in series; (c) an open groove cell to reduce reflection loss (Texas Instruments Research.)

needed to obtain a high density of good junctions, but promising performances have been reported.

11.2.2.   *Heterojunction cells*

We now turn to constructions which are less exotic although they use unfamiliar combinations of semiconductors. The reasons for considering these in place of silicon are their better optical absorption properties. With them we can get a better match to the solar spectrum and also have a higher absorption coefficient. By using two different materials in a heterojunction, it would be possible to extend the spectral response of a cell, albeit with the disadvantage of a more complex preparation procedure. Many of the semiconductors listed in table 11.2 have been prepared not only as single crystals but also as

TABLE 11.2.   *Semiconductor bandgaps (eV)*

	Bandgap E_g/eV	Direct or indirect	
$CuInSe_2$	1·01	D	
Si	1·12		I
Cu_xS	1·20		I
InP	1·34	D	
GaAs	1·43	D	
CdTe	1·44	D	
$CuInS_2$	1·55	D	
$GaAs_{0.78}P_{0.22}$	1·69	D	
CdSe	1·70	D	
$Ga_{1-x}Al_xAs$	$\begin{cases} 2\cdot20\text{--}1\cdot85 \\ 1\cdot85\text{--}1\cdot43 \end{cases}$	$(1>x>0\cdot34)$ D	I $(0\cdot34>x>0)$
GaP	2·25		I
CdS	2·42	D	
ZnSe	2·67	D	
$Zn_{0.24}Cd_{0.76}S$	2·50	D	
In_2O_3/SnO_2(ITO)	3·70		

less expensive thin polycrystalline films. Pairs of semiconductors which have the desired optical properties, and are also compatible for lattice match and electron affinity, are listed in table 11.3. If the electron affinities were to differ (figure 11.12) the conduction bands would not meet smoothly at the junction and there would be a 'spike' to impede the flow of electrons.

Some of the pairs in table 11.3 will be seen to have a narrower gap material on top of the wide-gap material, which is throwing away one advantage of the heterojunction structure. In these, photon absorption takes place close to the surface, and recombination losses are higher than for the alternative ordering. The problems of preparing

TABLE 11.3. *Semiconductor pairs for heterojunction cells.*

Top	Base	Theoretical efficiency (%)	Best efficiency (%)
p or n-$Ga_{1-x}Al_xAs$	n or p-GaAs	17·0	12
n-ZnSe	p-GaAs	15·6	13
p or n-GaP	n or p-Si	24·0	10
n-CdS	p-CdTe	17·0	8 (thin films)
p-Cu_xS	n-CdS	10·0	8 (thin films)
p-Cu_xS	n-$Zn_{0.24}Cd_{0.76}S$	15·0	4 (thin films)
p-InP	n-CdS	17·0	12·5
p-$CuInSe_2$	n-CdS		12 (thin films: 6·6)
p or n-GaP	n or p-GaAs	21·0	8
In_2O_3/SnO_2 (0·9:0·1)	p-Si	12·0	11
SnO_2	n-Si		10

the materials dictate this unsatisfactory arrangement.

The ternary compounds, such as $CuInSe_2$, $CuInS_2$, $CuGaSe_2$, provide a whole new range of semiconductors with suitable bandgaps for solar energy conversion, many having the required optical behaviour, but their technology has to be developed. The most promising cell in these compounds is perhaps the $CuInSe_2/CdS$ heterojunction, for this is compatible with thin film growth as well as the more usual single-crystal approach. Most of them give electrical-contact problems, and may develop a high series resistance during thermal processing.

The Cu_xS/CdS cell is more easily made than these cells, but has not been able to live up to the initial promise of a cheap, large-area, efficient cell. It is not a simple p–n diode, but is a p^+-i-n structure, as the copper diffuses into the CdS during manufacture and compensates the donors. Cuprous sulphide is a highly degenerate semiconductor and so the cell is very similar to a metal/n-type Schottky cell. p-Cu_xS is used in this cell because CdS can only be reliably doped n-type. Attempts to add acceptors are foiled by the development of vacant lattice sites which compensate them electrically.

In the fabrication of a polycrystalline CdS cell, CdS is deposited on a metallized substrate by condensation from the vapour in a

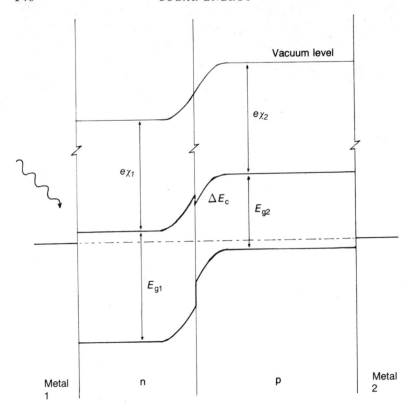

Figure 11.12. Heterojunction cell with an electron affinity mismatch, producing a spike in the conduction band.

vacuum-coating unit. After a light chemical etch, the film is dipped briefly in a controlled solution of cuprous ions to convert the top few hundred nm of CdS into the cuprous sulphide. Its composition changes after the next stage (and after ageing), going through a series of Cu_xS compounds, each with a slightly different bandgap and absorption coefficient. A thermal treatment of a few minutes in air at 200° C, followed by application of a fine-gridded contact complete the basic cell. Since the Cu_xS has a high resistivity despite the high carrier concentration, it must have a narrow grid spacing to ensure that the collection efficiency is acceptable. This grid may be only 80% transmitting and so offsets some of the good optical absorption of the copper sulphide. Figure 11.13 show the structure and appearance of a CdS cell. A final encapsulating film, is essential to protect the cell from atmospheric moisture and oxygen. Other sources of degradation do not need moisture but involve a photon-activated electrochemical plating-out of copper from the Cu_2S, if the voltage across the cell rises above a fairly low threshold.

A potentially cheaper cell can be manufactured by spraying chemical solutions on to a heated substrate, such as glass sheet, to form in turn a semitransparent electrode and the CdS. The chemical compounds decompose to give volatile side products which are easily removed from the film vicinity.

The band structure of the CdS cell is complex, and it has taken several years to arrive at a generally accepted explanation of its operation. Light is mainly absorbed in the Cu_xS, either during the first pass, or after reflection from the back contact. Some light of energy less than 2·4 eV is still absorbed in the CdS layer by excitation of electrons to copper impurity levels within the energy gap. This reduces the resistivity of the cell by a photoconductive effect, and produces a field across the top part of the CdS film. Indeed the voltage across a CdS cell practically all appears across this layer, and may explain the origin of the *cross-over* of the dark and illuminated current–voltage curves (figure 11.13(b)).

The CdS cell is still the most highly developed thin-film cell, but needs a longer lifetime, and efficiency higher than the present 8% if it is to be commercially viable, despite successful demonstrations of roof-top installations of these cells in the USA.

11.2.3. *Schottky barrier cells*

Another promising contender for cheap solar cells is the Schottky barrier structure, which may be thought of as a p–n junction with zero junction depth, and an optically attenuating metal film over the top surface (figure 11.14). Schottky cells are simpler to make than p–n silicon cells, use existing silicon technology, and require lower temperature processing. The problems associated with diffusion of dopants along grain boundaries in polycrystalline films are avoided by this technique and so thinner semiconductors may be usable.

All of these advantages are achieved without a great sacrifice in cell performance, for theoretical calculations of the efficiency arrive at a figure similar to that estimated for p–n cells. In practice it would seem to be very difficult to improve the AM1 efficiency of *true* Schottky barrier silicon cells beyond 10%, but even this is an economically acceptable figure. Although semiconductors other than silicon readily form Schottky diodes, and could be made into respectable solar cells without the need for a heterojunction structure, the pre-eminent position of silicon in the semiconductor industry and its abundance

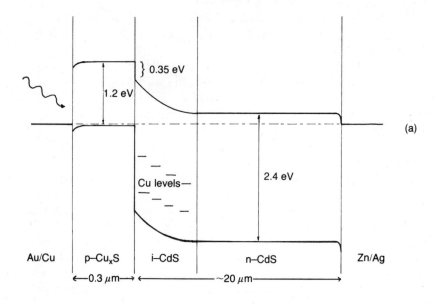

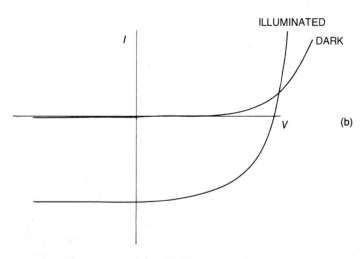

Figure 11.13.

make it difficult to beat. Nevertheless, GaAs Schottky cells have already given higher efficiencies (15%).

The simplicity of the Schottky diode, essentially a metal-semiconductor junction (figure 11.15) masks the complications which arise in *real* junctions through the existence of an interfacial insulat-

(c)

Figure 11.13. The CdS–Cu$_x$S solar cell. (a) Energy levels; (b) current–voltage characteristics. (c) Flexible CdS thin-film solar cell. (Reproduced by kind permission of IRD Co. Ltd, Newcastle-upon-Tyne, UK.)

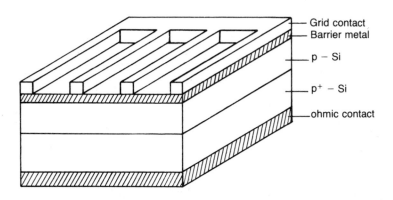

Figure 11.14. Schottky-barrier solar cell.

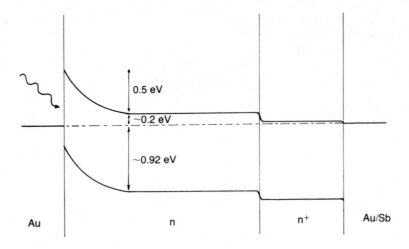

Figure 11.15. Energy levels in a typical Schottky diode.

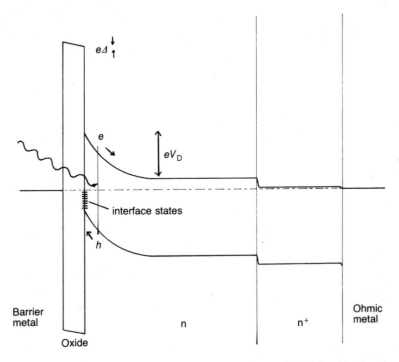

Figure 11.16. Energy levels for an MIS cell. This can give higher V_{oc} than in the absence of the I-layer (an insulating, or interface, layer, in this case oxide).

ing layer. This layer can reduce I_o without affecting the photocurrent. Figure 11.16 shows the energy band structure at the interface of an MIS (metal-insulator-semiconductor) cell. It is the controlled presence of this layer which allows high efficiency cells to be produced, but it also makes mass production of these cells more difficult.

An extended short-wavelength response relative to p–n cells (figure 11.17) should result from the elimination of both the surface 'dead layer' and the high front-surface recombination velocity, although it has been suggested that a very thin region may exist in the semiconductor in which carriers are drawn towards the surface instead of away from it.

Cell performance is also improved by adding a thin heavily doped region adjacent to the rear contact to reduce contact resistance, and to produce a BSF. More important an addition is a top-surface anti-reflection coating, for the high reflectivity of metals means that a large proportion of the incident energy is lost at the very beginning of the chain of events leading to electrical power generation. Not only is the theory of anti-reflection coatings on metals more complicated than for dielectric substrates, but the very thin metallic layers used in the solar cell are different from bulk metals and their optical constants must be obtained experimentally. The sheet resistivity of these thin metal films is a few tens of ohms per square, and the resistance can only be reduced by adding further metal, to the detriment of the optical transparency. (For an explanation of sheet resistivity, see Figure 11.18.)

Table 11.4 shows the achieved efficiencies for metal–semiconductor cells. If the efficiency of Si cells can be increased beyond its

TABLE 11.4.   *M(I)S cells*

Cell	V_{oc}/mV	AM1 I_{sc}/A m^{-2}	Efficiency (%)
Cr/Cu/Cr/p-Si	520	290	9·5
AR/Al/SiO$_2$/p-Si	470–560	260	8–10
Au/SiO$_2$/n-Si	550	220	9
Au/native oxide/n-GaAs	450	270	8·5
Au/oxide/n-GaAs	630	270	15
AR/Au/oxide/n-GaAs	630		15–20
AR/Au/native oxide/n-GaAs$_{0.78}$P$_{0.22}$			15

AR means a single quarter-wave anti-reflection coating.

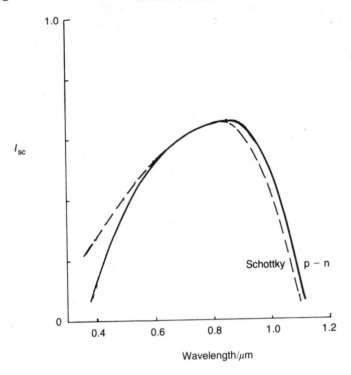

Figure 11.17. Comparison of the photocurrent spectral response for silicon p–n and Schottky cells.

present level to the expected 13% or more, and the control of the oxide thickness to <3 nm can be established on a production scale, these cells will make an appreciable advance towards the goal of cheap photovoltaic power. Real cost reductions await a thin-film version of the Schottky cell, and this may not be far away. Even higher efficiencies have been reached by GaAs and $GaAs_{1-x}P_x$ MIS diodes, in which the effect of a properly-tailored I-layer is even more dramatic. There still remains the difficulty of making a large-area device in which very thin layers of material must be maintained within narrow limits.

A further method of improving Schottky-type cells has been proposed: that of 'grating' cells. This technique has been applied to p–n silicon cells to reduce the area of the diffused layer and so reduce the junction leakage current. Figure 11.19 shows the appearance of one of these cells, made by alloying aluminium strips into n-type silicon to give thin p-type regions. As long as the lifetime in the *bulk*

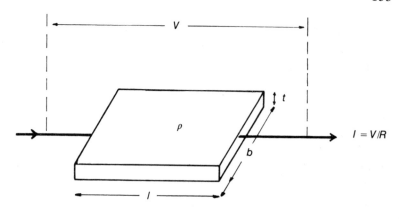

$$R = \rho \frac{l}{t.b} \, \Omega$$

$$R_{sheet} = \frac{\rho}{t} \, \Omega/\square$$

Figure 11.18. Definition of sheet resistivity.

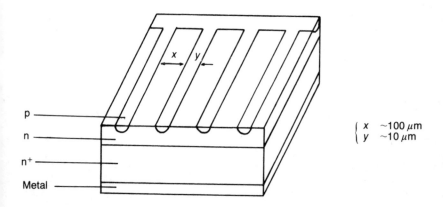

$$\begin{cases} x & \sim 100 \, \mu m \\ y & \sim 10 \, \mu m \end{cases}$$

Figure 11.19. The grating cell structure. Contact to the p-type strips is by the overlying metal strips.

n-silicon is long, carriers may be produced well away from the diffused stripes and still be collected by the junction. This means that the short-wavelength response is good, but that the longer-wavelength response is poor. A BSF would help to improve the collection efficiency of carriers from within the base. This structure has one problem in common with the normal p–n cells: its resistance is high unless the alloyed stripes are well contacted by metal stripes (either aligned or perpendicular).

Schottky cells using the same structure would not necessarily have lower leakage currents than conventional Schottky cells, since this is often determined more by recombination rates than by junction area. The stripe width and spacing should be optimized for any particular set of values for the diffusion length, lifetime, and transmittance of the metal. Whether this is a worthwhile exercise depends on whether the junction or the bulk is the limiting factor on collection efficiency.

11.2.4. *Thin-film cells*

Thin films of compound semiconductors may be produced by vacuum deposition: why is silicon not being made in this form, since it is otherwise a desirable material for cheap cells? Although silicon thin films have been made by vacuum evaporation from a crystalline

TABLE 11.5. *Thin-film solar cells.*

Film	Junction	Efficiency (%)	Date
Vacuum evaporated			
30 μm Si	p–n	1·5–3·5	1976
50 μm GaAs	Pt–MIS	5	1967
10 μm InP/CdS	p–n	2·8	1976
Chemically sprayed			
2 μm CdS/Cu$_x$S	p–n	5	1976
Vapour deposited			
10 μm CdTe/Cu$_x$Te	p–n	6	1967
CdS/CuInSe$_2$p–n		6·6	1976
CuInSe$_2$p–n		3	1977
CuInS$_2$	p–n	3·3	1976
Glow discharge decomposition— amorphous Si	p–i–n	2·4	1976
Glow discharge decomposition— amorphous Si	Pt–MS	5·5	1976
Glow discharge decomposition— amorphous Si	Ni/TiO$_x$MIS	4·8	1977
Chemical vapour deposition 15–30 μm n-GaAs	Au–MOS	6	1978
Chemical vapour deposition 25 μm Si (epitaxial on metallurgical Si)	n–p–p$^+$	10	1978

source, it is really necessary to have ultra-high-vacuum conditions for a reasonable degree of purity and perfection in even polycrystalline layers, for Si readily reacts with residual oxygen in a normal vacuum chamber. Ultra-high-vacuum chambers are not so readily adapted for industrial use since the attainment of a working vacuum below 10^{-8} Torr necessitates special techniques.

Cells made from most other thin-film semiconductors have not been much more successful than thin-film silicon cells, and still have a long way to go before they are sufficiently good to be commercially interesting. Most promise is perhaps shown by Cu_xS on CdS, and CdS on $CuInSe_2$ (table 11.3), but these have many problems to be solved yet. Table 11.5 shows typical quoted efficiencies of some thin-film cells, and is more an indication of the best achieved so far, than the performance to be expected of a large number of cells. From figure 11.20 it can be seen that the achieved efficiencies are not being

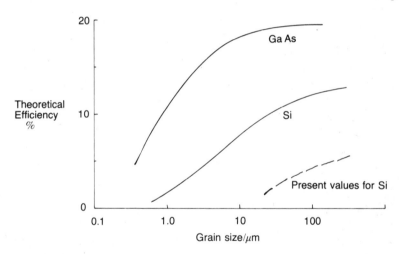

Figure 11.20. The dependence of solar-cell efficiency on the grain size of polycrystalline semiconductors (after H. J. Hovel).

limited directly by the small grain size of these polycrystalline layers, nor is the narrow depth of the layers a dominant factor (figure 11.9). Instead it is the high recombination in thin-film semiconductors which limits their usefulness in devices. Great efforts are being made to reduce this defect, which is caused principally by the grain boundaries.

Despite these efforts it is more likely that silicon-cell costs will be reduced by moving to another technology altogether. There are at least three new methods of preparing cheap silicon of a quality sufficient for solar cells, each at a research stage. Two of these involve drawing a long thin strip of crystal from a molten mass of the element, and have proved capable of giving silicon for 10% efficient cells (using conventional diffusion techniques) although the present yield of cells with this performance is low, and 5–6% is more typical. The growth of silicon ribbon by edge-defined, film-fed growth (EFG) has created a great deal of interest since its announcement in 1974. By drawing silicon through a die of high-purity graphite, a continuous ribbon of single-crystal silicon, 25 mm × 250 μm in cross-section, is extruded, and wound on a drum at rates up to 1 m h^{-1}. This eliminates the costly, labour-intensive stages in standard solar cell manufacture: growth of single-crystal boules from purified polycrystalline lumps, and subsequent slicing and polishing. The ribbon material is already in the form required for solar cells and so gives little waste. Its transport properties have been shown to be good enough for efficient carrier collection as long as the die is meticulously cleaned to remove lifetime-reducing impurities. Defects in the ribbon tend to be crystal twins appearing as striations parallel to the edges, and do not greatly affect cell behaviour. Figure 11.2 shows the EFG arrangement, together with other ribbon-growth methods.

A method which attempts to give ribbon without any die contamination is dendritic web growth. This is a simpler approach to the problem but is not as well controlled as EFG. Figure 11.22 shows the shape of a silicon strip pulled from the melt by inserting a seed crystal and allowing a 'button' of dendritic silicon to form before drawing it slowly away from the melt surface. The trick is to prepare the seed so that dendritic growth (i.e. like the formation of ice on windows) can occur. The dendrites then define the shape of the web which forms the majority of the ribbon. What makes control of this method difficult is the complex pattern of heat flow around the forming solid. p–n cells made with this silicon perform as well as vapour-phase grown silicon cells (Czochralski growth), i.e. over 10% efficient, and are potentially much cheaper. As yet it is only melt-grown single-crystal boules which are the source of silicon for commercial cells, on any reasonable scale.

Other laboratories around the world are trying to prepare sheets of good quality silicon by dip-coating a substrate, or by casting from a melt, or by deposition from the vapour on the surface of molten tin. A potentially more important contribution to large-area silicon growth has been made by a group at Dundee University, following leads from earlier work. They have been able to demonstrate that

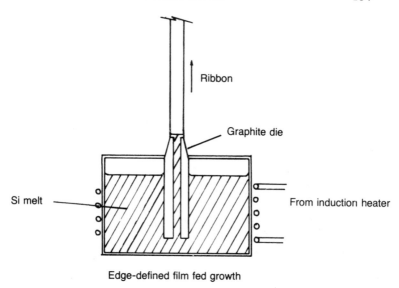

Edge-defined film fed growth

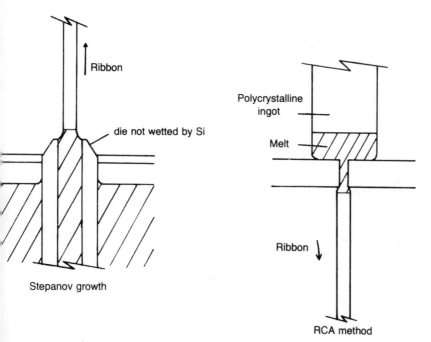

Figure 11.21. Ways of growing ribbon silicon.

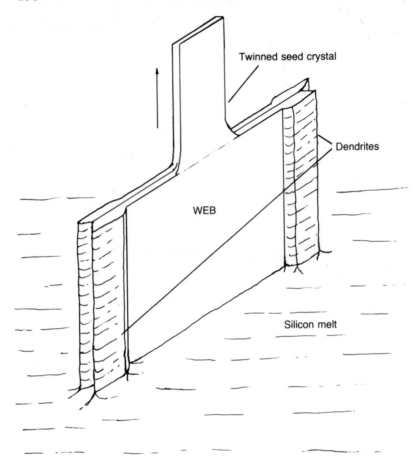

Figure 11.22. Growth of dentritic-web silicon.

amorphous silicon films can be doped both p- and n-type in a similar manner to crystalline silicon as long as the films are correctly prepared. To see why this is so, when other attempts to make certain silicon thin-films conducting have been unsuccessful, we must explain the meaning of 'amorphous'.

Crystalline silicon has a regular ordered array (lattice) of silicon atoms each tetrahedrally linked to nearest neighbours, with a constant chemical bond length and bond angle. Amorphous silicon has no long-range order in the atomic arrangement, but locally the bonds are probably mostly satisfied, with some variation in the bond angle although the bond length remains constant. This results in some

Figure 11.23. A model of a random network, like that of amorphous silicon, containing impurity atoms (e.g. hydrogen) on the dangling bonds.

atoms having more than the usual number of next-nearest neighbours, (see figure 11.23). There is really no reason why such a structure should *not* be semiconducting, since we showed in section 3.2.2 that the conduction and valence bands of a semiconductor arise from the overlap of individual energy levels from separate atoms brought together. There is no need for the atoms to have a long-range ordering. Of course, it is not to be expected that our amorphous solid will behave in every respect like a crystalline semiconductor. Since some of the atoms are not in the same local environment as would be the case in a regular lattice, they will contribute energy levels differing slightly from the norm. Those atoms having unsatisfied bonds (dangling bonds) will also give rise to energy levels away from the conduction and valence bands. Therefore, an amorphous solid of the type we are discussing will have a distribution of energy levels within the normally forbidden gap (figure 11.24). These states are not uniformly and densely distributed throughout the solid, but are localized. Movement of electrons between them is less easy than along the continuum of conduction band states.

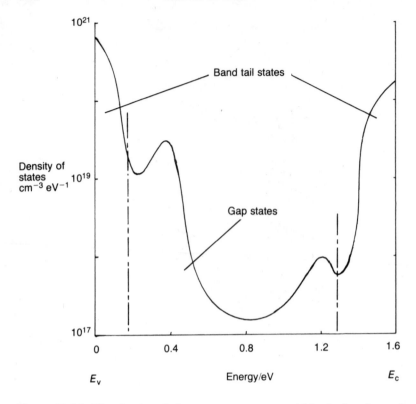

Figure 11.24. The density of electron energy states within the bandgap of *amorphous* silicon (prepared by glow discharge decomposition). (After W. E. Spear and P. G. LeComber.)

In most types of amorphous silicon the gap states are present in such numbers that any donors or acceptors can make no impression on the conductivity, for added charge carriers are quickly trapped. In the layers prepared by the Dundee group, it seems that hydrogen is incorporated into the lattice in a special way, to compensate the gap states, perhaps by taking up any dangling bonds in the film. Hydrogen is unavoidably produced in depositing their films, for the silicon source is the gaseous compound silane (SiH_4), which is decomposed by a glow discharge. Doping, by adding the gas phosphine (PH_3) or diborane (B_2H_6), is not as efficient as in crystalline silicon, for much is found to be electrically inactive in the films, but there is adequate conductivity control for solid-state devices.

The few semiconducting diodes made with this material have already surpassed some more complex and mature heterojunction technologies. The poor carrier lifetime in these films means that carriers generated more than 1 or 2 μm away from the diode junction will not be collected unless there is a large drift field. This necessitates the use of very thin cells, which is certainly sparing on material, but does not aid optical absorption. To some extent the loss in absorption is offset by the high absorption coefficient of amorphous silicon, together with its wider optical band gap (1·4–1·6 eV). Present efficiencies range up to about 6% (for an MIS cell), and efficiencies of 12–15% are expected for both p–n and MIS cells when the quality of amorphous silicon is improved.

With the long-term potential of amorphous semiconductors and the even longer term possibility of organic semiconductor cells some way from being realized, the solar cell industry is attempting to reduce the cost of more conventional cells by short-cuts in processing. Good performance cells have resulted from the use of silk-screened contacts (i.e. printed from a conducting ink and then heat-treated) rather than evaporated contacts, from the use of 'fired-on' paints for producing the dopant, and from aluminium rather than boron as an acceptor. Using low-purity metallurgical grade silicon has been less successful, but it is hoped to define the maximum impurity levels which can be tolerated in solar cells, rather than remove *all* impurities. Some solar panels can now pay back in less than 1 year the energy input during processing from the wafer stage, but to pay back the energy needed for producing the silicon wafers in the first place needs an additional three or more years, mainly because of the high temperature processing used in crystal growth and purification. Studies of the lowest energy route to solar-cell manufacture, starting from silica, show that solar energy could be used to produce solar cells without an additional energy source.

It is believed that thin-film solar cells offer the only prospect of economic solar electrical power, but present solar-cell applications are limited to special situations in which cost is not the major factor.

12. Biological conversion

12.1. *Photosynthesis*

Man already relies on solar energy for his basic needs. Photosynthesis is the natural conversion process which supports all life on Earth, through the chemical, chlorophyll. Although the net reaction involved is just the combination of carbon dioxide and water to yield carbohydrates (equation (12.1)), the discrete steps are not yet well understood. Glucose, $C_6H_{12}O_6$, and sucrose $C_{11}H_{22}O_{12}$, are examples of more complex carbohydrates based on this fundamental unit. No-one has yet managed to accomplish artificial photosynthesis *in vitro*.

$$CO_2 + H_2O \rightarrow CH_2O + O_2 \tag{12.1}$$

Figure 12.1 shows the important steps in this process. Two photons

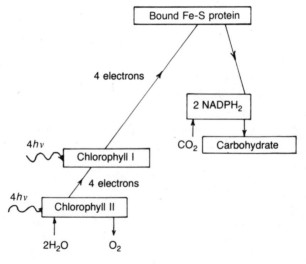

Figure 12.1. The important steps in the photosynthetic production of carbohydrate.

are required for each of the four electrons transferred, because photons of visible radiation (i.e. at the peak of the solar spectrum) have insufficient energy for a single-step reaction. Chlorophyll is green because it absorbs energy in the red and blue parts of the spectrum, although these are not the strongest parts of the Sun's output (figure 12.2). Some plants have additional pigments (such as

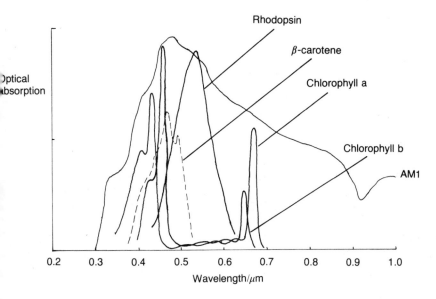

Figure 12.2. The absorption spectrum of some pigments, shown against the AM1 spectrum.

carotenoids) which *are* able to absorb energy in the green part of the spectrum and these aid photosynthetic efficiency. In contrast, the human eye responds most efficiently near the peak of the solar spectrum (555 nm), but the pigment involved, rhodopsin, is excited by single-photon absorption.

The result of fixing *one* molecule of CO_2 as sugar or starch by this absorption of *eight* photons is that only one third of the input energy is actually stored. This means that photosynthesis is at best 14% efficient at converting solar energy. The world mean conversion efficiency is about 1%, so there is room for improvement even within the restrictions of normal photosynthesis. Of course, plants growing in natural surroundings must compete for the other chemicals they require for continued growth, and indeed they consume part of their stored energy each night. These limitations, together with the ravages of pests, disease and weather, will make it difficult for field-crops to be produced much more efficiently.

It has been suggested that even with present-day yields a suitable crop could provide all of the UK food requirements together with 15% of our fuel, without increasing the present area of arable land. It is questionable whether the plant requirements additional to sunlight would continue to be met, and possible that a change in living standards would result from such an economy.

12.2. *Improvement of crop yields*

Apart from water and CO_2, plants must have nitrogen to manufacture sugars. Inert *atmospheric* nitrogen cannot be used, and instead a more active form such as ammonia must be provided. Man can produce this chemical, using a high energy input process, giving the common fertilizers. Certain bacteria (of the genus *Rhizobium*) can also 'fix' nitrogen by their enzyme, nitrogenase, but this only works in the absence of oxygen. Thus these bacteria enter a symbiotic relationship with leguminous plants (e.g. peas, beans) and form nodules on the roots within which nitrogen is exchanged for plant carbohydrate. By using crop rotation, nitrogenous material left from legumes can help to fertilize a different crop the following year. Attempts are being made to transfer this property to cereals by cross-breeding with certain tropical grasses which also associate with nitrogen-fixing bacteria. Genetic engineering may make it possible to provide a cereal crop with its own nitrogenase and anti-oxidant system.

Even without such advanced techniques, it is possible to reduce the amount of artificial nitrogen-fertilizers needed, by recycling nitrogenous material after a harvest. This means making greater use of sewage and animal slurry, and fermenting chemical-fuel crops rather than burning them, so that a nitrogen-rich residue is retained. Some fuel crops may even be grown on effluent to help in its purification. Water hyacinth can grow nearly 0.2 kg dry material per m^2 per day, and if the high water content can be removed before transport, it could provide a valuable energy crop.

Plant yields also increase in a CO_2-enriched atmosphere, and with an increase in temperature. Greenhouses provide controlled environments where such improvements may be made, but only with considerable expense. Greenhouse design may be improved even beyond the high standards of the nineteenth century, with modern thermal insulation and automatic ventilation. Solar- and wind-energy have been examined as possible cheap sources of heating, and industrial waste heat has also been used.

12.3. *Fuels from crops*

Growing plants is the only readily available method for converting solar energy to liquid fuels. Alcohol or methane is readily made from fermented crops such as potatoes, and have been used as fuels from such a source. Since water makes up over half of all plant material, it must be removed from a fuel crop. Evaporation of water requires over 2 MJ kg^{-1}, which is a significant part of the energy which it is hoped to extract from the crop. Therefore, one can either grow a sun-dried crop (e.g. cereals), or use fermentation for which drying is unnecessary. This discriminates against direct combustion.

Carbohydrate is the most valuable fuel constituent of plants, with a calorific value of about 16 MJ kg^{-1}. Most of this is in the form of cellulose, a polymer of sugar, and difficult to break down. Ruminants can eat grass because their digestive track harbours bacteria which can do this for them. The enzyme responsible, cellulase, can be extracted and will produce glucose from cellulose waste with about 50% efficiency. The glucose syrup may be fermented, or used as feedstock for protein production from micro-organisms.

Protein itself represents a large amount of stored energy, and should be returned to the land if not used as food. Chemical fuel crops should be selected for low protein content rather than waste it. Free species which resprout after pollarding are suitable, *and* may be grown on marginal arable land.

Hydrocarbons are more familiar fuels, but producing them by chemical means from crops does destroy protein. Alternatively, some plants produce hydrocarbons directly: latex from rubber plants is one example. Other hydrocarbons might be produced from plantations of related species.

Hydrogen has already been suggested as a fuel with wider prospects. It may be produced by bacteria such as bacteriorhodopsin using the enzyme, hydrogenase, sometimes competing with nitrogen-fixing bacteria. Bacteriorhodopsin matches the solar spectrum well, because it contains rhodopsin (figure 12.2). Hydrogen production from plant systems has not yet been demonstrated as a feasible technology.

It seems that biological conversion of solar energy has no new breakthrough of efficiency to demonstrate, but it will continue to be important for fuel production on a small scale.

13. Applications of solar energy

13.1. *Solar heating*

The greatest impact on new solar energy converters has been for domestic water heating. In favourable climates thousands of units have been installed, and even some factories have been able to provide low-grade process heat (such as heated water for food industries) sufficiently reliably by solar collectors. In the UK, about one third of all energy consumed is used for water heating and space heating, but despite possible energy savings for the country, it is not yet attractive to an individual to install solar panels unless these are fitted to the house as it is built. Not only do we have cheap gas for heating purposes, but the panels are not produced in sufficient quantities to benefit from mass-production price savings. Even if the lifetime of a commercially installed solar heater exceeds 20 years, the savings in fuel costs will not repay the investment for at least 7 years, and yet it *is* quite feasible to provide 50% of domestic heat by solar energy in many parts of the country. Owner-installed panels will be cheaper, but in order to achieve 20 trouble-free years of operation, all plumbing and construction must be of a high standard. It is extremely probable that the circulating pump would need replacing within this period, and quite possible that pipes or storage tank would suffer corrosion, reducing the efficiency of the system. A *small* panel might be a better buy at this stage in their development, to provide at least summer hot water without the problems of storing energy for sunless periods. It is unrealistic to speculate on how costs of solar collectors will fall in future years, for when conventional energy rises in prices, then so does the cost of materials, transport, and everything else in the manufacturing chain. Widespread adoption of solar panels in the UK would be stimulated by a tax incentive scheme.

Several autarkic houses have been built in recent years, designed to be self-sufficient in energy. These use a combination of solar, wind and bio-energy, together with elaborate energy-conserving techniques, such as a heat-pump to extract energy from outgoing water and air. A heat-pump is a large capital investment, but one which would be more widely made if its features were publicised and a domestic-sized unit was available. Heat-pumps are refrigerators 'working backwards' which extract heat from *outside* the building (the ground, or air, or a water supply) and deliver it at a higher temperature inside. The coefficient of performance (heat removed from the lower temperature source, divided by the net work done during the refrigeration cycle) should be at least 3 for a useful gain. Since heat-pump efficiency is ultimately limited by the Carnot efficiency, solar collectors can be used as the heat source rather than a colder source, and will be more effective in this way when insolation is weak.

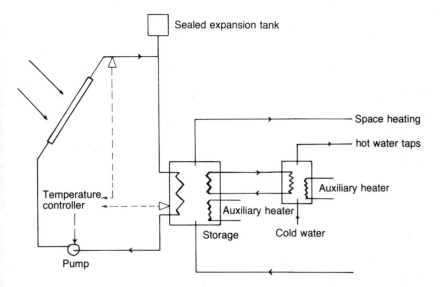

Figure 13.1. A solar heating system for providing hot water and space heating.

A space- and water-heating system is shown schematically in figure 13.1. The addition of thermostats and valves to control the input of solar or conventional energy (e.g. off-peak electricity) and to prevent the system going into reverse and radiating heat to the sky during the night, is essential if the system is to live up to its promised perfor-mance.

A simpler scheme to operate, but one which is further removed from conventional architecture is the passive system used by the annexe to St George's school, Wallasey, opened 15 years ago. Solar collection is provided by a vertical south-facing wall which is heavily glazed, and behind which is a heavy concrete structural-support wall. Generally, in a *house* provided with space- and water-heating from solar energy, the solar collectors need up to two thirds of the floor area in order to provide a reasonable proportion of the required space-heating, plus a further 1 m^2 per person for hot water. In addition, an energy store, commonly using water (about 50 kg/m^2 of floor area) or pebbles (about $0 \cdot 2$ m^3/m^2 of floor area), must be provided. The Wallasey school combines these requirements in the solar wall and roof slab, and the combined heat from the sun, the occupants, and the tungsten light bulbs ensure that no additional heat source is needed, even in winter. Indeed, in summer, ventilators must be used to prevent air temperatures rising uncomfortably, since no screening was provided for the solar wall. A feature in some dwellings built on this principle in the USA is the large overhanging eaves which shade the windows from the midday summer sun whilst allowing full exposure of the low-altitude winter sun. The Wallasey school relies on free convection to draw air through vents near the bottom of the concrete solar wall, across the gap between this and the glazing, and back through vents at the top into the classrooms behind. In some ways this use of heavy masonry to store solar heat reverts to traditional building methods: an insulated roof, and walls with only small windows. A drawback of the Wallasey solar wall is its large heat loss, although the installation of movable screens would reduce this. (Heat loss through the other walls is prevented by heavy thermal insulation on the *outside*.)

On a less elaborate scale there are several important uses of solar energy for which a relatively simple construction will give good performance, although once again the most efficient and maintenance-free devices require greater expertise. Such applications include those for which the provision of heated air or water would extend the periods during which solar energy is already in use: swimming pools, greenhouses, fish farms, and crop drying. Outdoor swimming pools are very expensive to heat by fossil fuel boilers, but even a simple, cheap, flat-plate collector can provide solar heating for raising the water temperature by a few degrees. Since the temperature rise across the collector need not be great, there is no need to add a cover glass to prevent heat losses. On the other hand, a cover over the pool itself would prevent significant energy loss which must otherwise be made up by a larger solar collector area. Similar collectors can be used to heat the water in enclosed tanks for fish-farming, in order to extend the breeding season and growth periods. Crop drying in many parts of the world relies heavily on

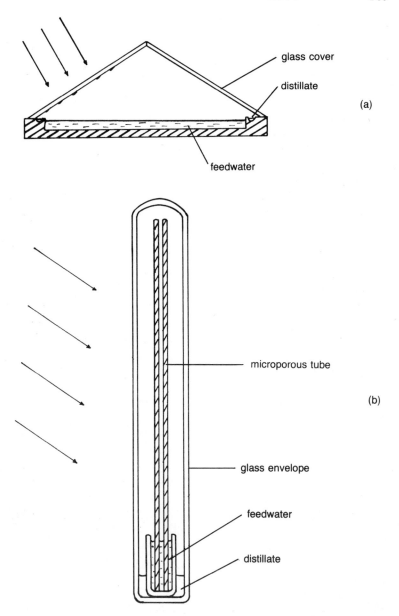

Figure 13.2. Solar stills for impure water. (*a*) Conventional 'greenhouse' pattern still; (*b*) still using polyethylene wick. (Reproduced by kind permission of J. P. Coffey.)

uninterrupted periods of direct sunshine, but drying of grain or fruit on the plant or on open-drying floors leads to much wastage. Since drying without spoilage must use very high temperatures or only low temperatures, a simple solar collector can provide controlled warm-air which is blown through a silo by fan. A blackened surface enclosed with a transparent polythene tube is perfectly adequate for the purpose, without involving a large capital investment. Alternatively, solar heating could be used to recharge a desiccant in a drying tower when it has become saturated with extracted moisture from the crop.

A problem which can arise when damp air is heated in a solar collector is condensation on the inner surface of the window. This suggests an application in which hot water or air is not the required product; i.e. distillation of brackish or salty water. Water passing slowly along open channels in a covered solar collector will be heated sufficiently for some to evaporate, and condense on the cool glass cover. If the cover is tilted the condensate will run down for collection in a channel at the lower edge (figure 13.2(a)). Plastic covers are not very satisfactory, for apart from degrading more rapidly than glass they often mist-up rather than forming larger water droplets which can run off. The consequent reduction in solar transmittance of the cover reduces the effectiveness of the still. Although use of the Sun's heat to produce salt from brine is an ancient industry, use of the evaporated water does not appear to be as old. One of the oldest large-scale stills was one installed in Chile in the 1870s to produce drinking water for mules at nitrate workings, from well-water containing 14% dissolved salts. The 5580 m² glass-covered still provided about 19 000 litre/day for over 10 years, until a railway replaced mule transport. More recently, large stills have been installed on Greek islands following the development of an 8600 m² still built on Patmos in 1967. A novel design proposed by J. P. Coffey in the USA employs a tubular 'wick' of blackened microporous polythene enclosed within a transparent tube (figure 13.2(b)). Impure water is drawn up the wick, from which it evaporates, leaving the impurities to flow back down the surface of the polythene. Water condenses on the cover tube and runs down to the base for collection. This still might be floated on the sea, or stuck into the ground to extract fresh water. Yields have been encouraging, but overall efficiency could be improved further by using regenerative feedback to preheat the intake from the latent heat of condensation. In this 'multiple effect' still, not all of the absorbed energy is rejected on condensation, and less energy per litre of pure water is required.

Let us now turn to applications requiring higher temperatures than those reached by simple flat-plate collectors. The advantage of solar furnaces lies not so much in the high temperature at the focus but rather in the speed with which the heat flux can be changed. Although solar steam boilers were shown at the 1878 Paris exhibition, they remain too costly for widespread use. A much smaller parabolic reflector, of 1 m² aperture, is all that is required for cooking food, or sterilizing medical instruments. Solar cookers for home barbecues are available in various designs in the USA, but their everday use (even in an ideal climate) necessitates a change in social customs! It would be easiest to cook a meal at midday, but especially in hot climates it is the custom to eat in the cooler evenings. Solar cookers have been introduced to India in large numbers (figure 13.3), in an effort to reduce the use of dung as a fuel, but habits take many years to change. An interesting design for a solar cooker which does not need to track the Sun, and which allows the cooking pots to be placed within a dwelling, uses a modified flat-plate collector to produce steam which is then fed along a pipe to an insulated cooking chest. A truncated conical reflector with an axial absorber has been designed and tested for the sterilizing of medical instruments used by field doctors in Iran. The absorber consists of two concentric cylinders, the space between which contains water, and the inner of which contains the instruments. A satisfactory temperature for a wet steam

Figure 13.3. Solar cooker, developed in India, being demonstrated at the National Physical Laboratory, New Delhi.

autoclave is above 120° C: 6 minutes at 126° C will thoroughly sterilize the contents.

13.2. *Solar cooling*

Of more importance in many ways than solar heating is the use of solar energy during very hot weather for air conditioning and food preservation. There is no practical reason why solar electrical power should not be used to drive a version of the domestic refrigerator, but this requires high collector temperatures or expensive solar cells. Only absorption cooling has attracted much attention from modern solar engineers, but of great interest and historical importance is radiative cooling (the unwanted reverse action which can occur with flat-plate collectors at night if they are not fitted with non-return valves). The absorption refrigeration cycle is compared with the compression system in figure 13.4.

For *intentional* radiative cooling the 'collector' should be designed to *release* heat as long-wavelength radiation to the sky. Although convection was an important path for the emission of energy from our flat-plate heaters, it does not follow that convection currents are desirable in a radiative cooler. In fact, if we rework the energy balance equations used for the flat-plate collectors, for the system depicted in figure 13.5(*a*), it is seen that conduction/convection energy transfer is in the direction of *supplying* heat to the absorber. The infra-red transparent *screen* is, therefore, an essential feature to

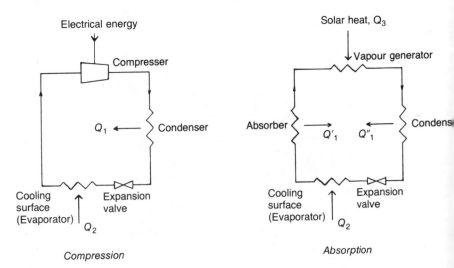

Figure 13.4. Comparison of refrigeration cycles.

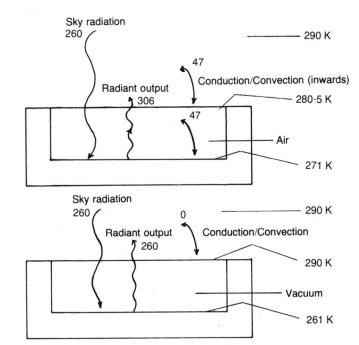

Figure 13.5. Radiative cooling through an infra-red transparent cover. (Energy fluxes in W m^{-2}.)

prevent heating. Notice that the screen itself cools to nearly 10 K below ambient. Since the screen is transparent to the emitted radiation, it will also allow in the long-wavelength sky radiation. This may be weak in dry, clear conditions, but in a humid climate it would be better if a selective transmitter could be used to block the sky radiation. This is mostly emitted by CO_2 and water vapour in narrow spectral bands, leaving a window between 8 and 13 μm, in which region there is little sky radiation. A selective transmitter (figure 13.6) could be chosen to radiate outwards through this window, even during daylight if it was screened from the Sun.

The actual cooling expected from our radiator with a *non-selective* cover is shown in figure 13.5(*a*) to give an *isolated* plate temperature of 271 K, even when the sky temperature is taken to be above 273 K. This is because the thermal input is too low to keep up with the radiated output. A suitable radiator plate could be made from glass, for this has a high infra-red emissivity, and a low absorptance for visible radiation. If we could have a vacuum space between the plate and cover (figure 13.5(*b*)), the further reduction in thermal input produces a plate temperature of 261 K. The cover temperature in this

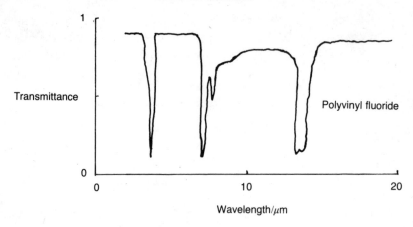

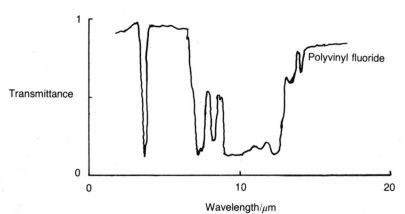

Figure 13.6. Optical properties of plastics suitable for use in radiative coolers. Polyvinyl fluoride is a good radiative emitter in the 8–13 μm band, but a poor absorber over other wavelengths. It is therefore suitable for coating a metallic radiating surface. Polythene is a suitable infra-red-transparent cover.

case is higher, which helps to prevent the formation of dew, although in practice it may be a difficult structure to make. (Water is, of course, opaque and scattering to infra-red, and so the plate would not cool as well as before, as indeed would be the case for a real cover with a finite infra-red absorptance.) As with the flat-plate collector, the actual plate temperature in an operating system would depend on the flow rate of the liquid or gas across it, and on the thermal constants of the materials. Radiative coolers of a simple construction have been in use for many years in some parts of the world, to make ice in shallow radiatively cooled ponds. These ponds are screened from the Sun by a high wall to the south, and lose heat during the clear nights to such an extent that ice forms (e.g. the 'yakhchals' of Iran).

Note that it is easier to store cold water than hot, without thermal leakage, and so production of chilled water at night for use during the day is perhaps easier than the storage of hot water for use at night. The high-efficiency radiative cooler awaits new stable materials with the desirable spectral properties outlined here. In fact, it is not essential for temperatures below 0° C to be reached in order to store fresh foodstuffs for short periods, and the temperatures calculated for isolated radiating plates indicate that useful amounts of cooling power could already be obtained in a working system.

13.3. *Solar electricity*

We have seen that an early use of solar heat-engines was in the provision of power for irrigation pumps; now we are entering a period in which photovoltaic cells are becoming competitive not only with these but also with diesel generators. In many countries it is acceptable to have irrigation water delivered only when the Sun is shining, rather than on demand, and energy storage for other periods need not be considered. Solar cell irrigation is receiving great attention, and if cell costs fall to the $2 per peak watt expected by some observers for 1980–2, then many irrigation schemes will become worthwhile, for not only must one weigh the relative installation costs, but also the maintenance needed, the operating costs, and the lifetime of the systems. A recent installation in Nebraska, USA, uses 97 000 cells to generate 25 kW peak, in combination with 80 kWh of storage-battery capacity, and has powered a d.c. pump motor for cornfield irrigation. Although 'big business' agriculture in the States is used to paying for expensive irrigation, these cells at $5 each are still too costly. At a future price of $2 per peak watt, 100 kW arrays become of interest for supplying remote villages with power. Note that the entire production of silicon cells in 1975 was only about 100 kW worth, and less than 1 MW worth in 1977.

At present, solar cells are used in situations where their reliability offsets the cost of maintenance visits: remote radio beacons, telephone repeater stations, VHF radio links, navigation lights, remote instrumentation, cathodic protection of pipelines. These applications have often tested cell modules under severe environments and it has usually been the encapsulation rather than the cells which have suffered. Plastics have been particularly prone to attack and must be improved if intrinsic cell lifetimes of 15–20 years are to be attained by complete panels.

Some applications are particularly suited to solar-cell power because they have a short-duty cycle: limited periods of operation mean that reduced storage-battery capacity is acceptable. One such case is the provision of educational TV in West African villages. The sets require about 32 W at 30 V d.c., and are in use for 30 h/week. The storage batteries, diode-protected against discharge through the cells overnight, and equipped with regulators to prevent overcharging, are lead–acid accumulators of 40-A h capacity. Cells are mounted above school roofs, allowing air to circulate beneath to avoid overheating, and provide 1·1 kWh/week. It should be noted that large arrays require protection against lightning strike. Developments from field tests such as these include solar-cell powered refrigerators for hospitals in remote villages, perhaps using concentrator arrays, for these are expected to be cheaper in the period before thin-film cells are fully developed.

As one can well appreciate, a serious problem of generating electricity from the Sun is in providing adequate storage to cope with inclement periods and hours of darkness. Even if very cheap cells are developed, this high cost for electricity storage weighs strongly against conventional battery use: solar electricity generated on the Earth's surface is still uneconomic when compared with any other generation route.

An alternative idea which avoids the need for storage to cope with seasonal, diurnal, and climatic variations of insolation is the power satellite proposed by P. E. Glaser (Figure 13.7). By fixing a large array of cells in geostationary orbit, 22 300 miles above the Earth, they will be exposed to the Sun all year, apart from brief periods of eclipse at the equinoxes. Other advantages become apparent when the scheme is studied in more detail: for example, the absorbed solar energy rejected as waste heat would no longer be fed into the Earth's atmosphere but would be radiated into Space. The supporting structure could be less dense than needed on Earth, and would not need to be a series of inclined planes taking up valuable ground space. However, thermal stresses during the brief eclipse periods would be severe, and pose a difficult engineering problem. It is

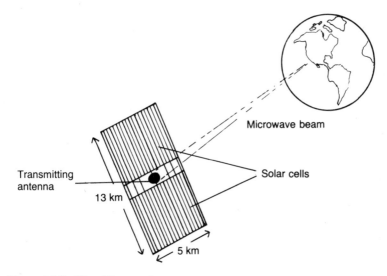

Figure 13.7. The Glaser solar-power satellite. Energy is to be beamed to Earth as microwave radiation.

proposed that the collected solar energy would be converted to 10 cm wavelength microwave radiation and beamed to Earth from a 1 km diameter transmitting antenna. At this wavelength there would be negligible atmospheric absorption or refraction, and no atmospheric ionization. The generation of microwave power from solar cells should be 90% efficient, and transmission through Space is almost lossless. Subsequent collection and rectification at a 10 km diameter receiving antenna (which would be 80% transparent to sunlight) would be 90% efficient, giving perhaps 77% overall performance for the complete power station. (Some power would be used to maintain a stable orbit, and to steer the transmitting antenna.) The major cost would be the transport of materials and parts to the orbiting site (making up nearly half of the $1500 per installed kW), but it is possible that cells could be manufactured in space from lunar material at an overall energy saving. Present 15% efficient Si cells have a mass-to-power ratio of 14 kg kW^{-1}, but substantial savings should be possible by combining concentrators with thin-film cells. A satellite lifetime of at least 30 years is proposed, during which cells would have to be successively replaced to overcome degradation caused by radiation damage (a problem already familar to satellite scientists). An alternative conversion technique using Brayton-cycle heat-engines to drive turbogenerators poses even more problems.

This concept is far beyond any application of solar energy now being put into operation, but it would be capable of providing a great deal of the World's future energy needs if a large number of satellites was constructed. It is questionable whether such a large committment of our resources will be made, or should be made, for withdrawal from the scheme, once started, would be difficult.

14. Energy storage

14.1. *Thermal energy storage*

There are few applications of solar energy which do not require either a back-up power supply or some form of energy store. We have already mentioned the obvious use of sensible-heat storage in a tank of water or in a pebble bed, and this may be improved by using latent-heat storage. The advantage of storing energy in a reversible phase change of a medium (usually a solid–liquid transition) is that the store is at a lower temperature than the equivalent sensible-heat store, and so the thermal losses are reduced. In addition, the store remains at a more constant temperature, which helps to maintain a constant rate of extraction of heat. Latent-heat stores have a lower volume and smaller surface area than does the equivalent water-tank store, and so their rate of loss of heat to the surroundings is less anyway.

Glauber's salt, $NaSO_4 . 10H_2O$, has been more thoroughly studied than any other of these materials. It dissolves in its water of crystallization at $32.4°$ C, with a latent heat of 250 kJ kg^{-1} (370 MJ m^{-3}), which matches the output temperature of a flat-plate collector very closely. Unfortunately, the anhydrous salt is rather insoluble, and slowly precipitates during several phase-transition cycles. The effect of this incongruent melting is an apparent drop in the latent heat as the salt fails to recrystallize fully. Mechanical stirring, additives and ultrasonic vibration have failed to overcome this flaw. An alternative salt, $Ca(NO_3)_2 . 4H_2O$, can store 144 kJ kg^{-1} (259 MJ m^{-3}) at $47°$ C but suffers from supercooling: it tends to remain liquid below the thermodynamic freezing point. Nucleating agents can help to prevent this, but many other salts which have yet be tested may suffer no such defect. Organic compounds, such as waxes, have lower latent-heats (around 150 MJ m^{-3}), undergo large volume changes on melting, and have poor thermal conductivity. Acetamide (CH_3CONH_2) forms a number of compounds and eutectics with other substances, giving a range of storage media melting at 35–60° C, with specific latent heats around 200 kJ kg^{-1}.

The large photothermal power plants may require a high-temperature store, although rock beds are quite suitable for this purpose. Some of the fluorides have large latent heats and offer a higher storage density. A lithium fluoride–sodium fluoride eutectic melts at 652° C (800 kJ kg^{-1}).

In all of these systems the exchange of heat is localized at the solid–liquid boundary, which moves through the container as heat is added or extracted. A floating heat-exchanger is likely to become 'frozen-in' during heat extraction, unless the chemicals are packaged in small capsules. Some of the advantages of latent-heat storage are thus countered by the extra expense of ancillary equipment.

14.2. *Mechanical energy storage*

Gravitational potential energy is the only storage cycle currently used by the electrical generating boards. This is an extremely efficient method, although the storage density (1000 kg of water pumped a few hundred metres stores only 3·6 MJ) is low. Typically, cheap night-time electrical energy is used to pump water to a high reservoir, to cope with surges in electricity demand during succeeding days, by using turbine generators. Thus power stations can be operated continuously at near full-power. Current pumped-storage capacity in the UK is approaching 2·5 GW, which is only about 5% of peak electrical power demand, and the number of sites which can be used is diminishing.

Kinetic energy storage by flywheel is familiar on a small-scale in mechanical machinery, but is now being considered for large-scale energy storage. It is possible to design almost frictionless bearings, with long life, using modern materials, but the possibility of large flywheels breaking up under stress is a great danger. A steel flywheel could store up to 180 kJ kg^{-1} without exceeding its tensile strength, and carbon fibre or fused silica would be even better.

Compressed air is another possible energy store with an electrical compressor input, and a turbine-generator output. Loss of efficiency arises from the increase in air temperature during compression, which also increases the volume to be stored. Much of the compressor input energy is wasted as heat in this way, unless a regenerative heat store is added to cool the air before storage, and heat it afterwards on its way to the turbines.

All of these methods are suited more to large-scale applications, than to the distributed power units which are better for solar-energy conversion.

14.3. *Chemical energy storage*

Chemical energy storage provides the most compact and efficient method, but is also the most expensive. One of the most pressing needs is a storage cell to link with cheap photovoltaic cells, but even the involvement of car manufacturers in this field has failed to solve the problem. Lead–acid batteries suffer from high cost, short lifetime, and comparatively low energy-density ($600 \, kJ \, kg^{-1}$), but have proved difficult to beat. A solar-energy system requires a much longer lifetime of its storage medium than the few hundred charge–discharge cycles of this battery, but suitable alternatives are more expensive. Lithium–chlorine, sodium–sulphur, and zinc–air cells are all under development, and appear promising.

Since hydrogen has been proposed as a chemical fuel, derived from the electrolysis of water, it might be combined directly with oxygen in a fuel cell to generate electricity. Laboratory cells have achieved 60% conversion efficiency, but have yet to prove sufficiently reliable for long-term operation.

15. Economic and technological prospects

This book has attempted to show that various solar-energy converters already exist in an advanced state of development, but have not been adopted since conventional energy sources are still much cheaper in most parts of the world. The economic value of solar energy depends on whether one examines it as an individual or from the point of view of society. Since the savings in fuel costs obtained by substituting a solar water heater for a conventional boiler are low, a better financial return can be had by investing the money otherwise used for installation, and using the profit to pay for the more expensive fuel. This may not be in the best interests of the nation. Since the future cost of fuel is likely to increase more rapidly than the cost of solar-heating panels, solar heaters installed *now* may repay their cost more rapidly towards the end of their lifetime. This suggests that expensive, reliable panels would be a better buy than cheaper ones requiring earlier replacement. In either case, it is not feasible in the UK to expect to meet the whole annual demand for heating from simple solar collectors, and this will save the individual from installing energy storage any larger than that needed for diurnal storage. On the other hand, a well constructed building incorporating a great deal of insulation and heat-conserving devices will gain much more from solar collectors than a conventional house does, and should be an important investment for the future.

Photovoltaic cells can already give the performance and lifetime required by many applications, but these parameters for the cheaper thin-film cells are not yet adequate. It is thought by many that concentrator cells using gallium arsenide are already viable for large-scale use, in climates with long periods of unobscured sunshine. Silicon cells with low concentration of sunlight are also very close to being economically acceptable. A maximum acceptable price for large silicon-cell arrays is around $1 per peak watt, which is about one tenth of current prices, and a short-term goal in the USA. Large-scale solar-electricity generation in the UK will need to supply power at a cost of a few pence per kWh, compared with 50–150 pence/kWh at present, to be competitive with conventional sources. Thin-film silicon, and especially amorphous silicon, cells will need to reach a minimum efficiency of 8% before they are viable, for the cost of a supporting structure, encapsulation, and interconnection would otherwise be too great. If their initial promise is realized, solar-cell power stations of more than a few hundred kW capacity will be feasible. It is not possible to say if power stations on the scale of current conventional stations will be realistic, and it is felt that solar power is more suited to a *distributed* network of generators than to a central power station and grid system.

Despite the problems and limitations of solar-energy conversion, funding for the development of better collectors and converters is being increased in many countries. Solar energy *is* a non-polluting resource capable of directly supplying an appreciable fraction of the World's energy needs, and should not be treated as a soft-technology impractical alternative to nuclear power.

Appendix A. Heat engines and thermodynamics

The use of thermodynamics in understanding the operation of heat engines is discussed systematically in such books as Zemansky's *Heat and Thermodynamics*. These brief notes are intended only as a reminder for those already familiar with the subject.

The P–V diagrams of Carnot, ideal Stirling and ideal Rankine cycles are shown in figure A.1. Heat Q_H is added at temperature T_H,

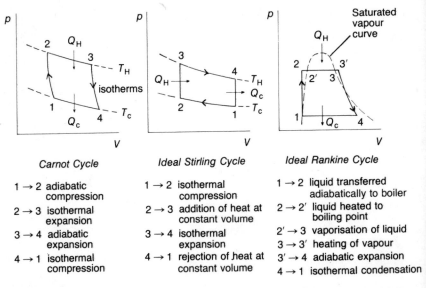

Carnot Cycle	Ideal Stirling Cycle	Ideal Rankine Cycle
$1 \rightarrow 2$ adiabatic compression	$1 \rightarrow 2$ isothermal compression	$1 \rightarrow 2$ liquid transferred adiabatically to boiler
$2 \rightarrow 3$ isothermal expansion	$2 \rightarrow 3$ addition of heat at constant volume	$2 \rightarrow 2'$ liquid heated to boiling point
$3 \rightarrow 4$ adiabatic expansion	$3 \rightarrow 4$ isothermal expansion	$2' \rightarrow 3$ vaporisation of liquid
$4 \rightarrow 1$ isothermal compression	$4 \rightarrow 1$ rejection of heat at constant volume	$3 \rightarrow 3'$ heating of vapour
		$3' \rightarrow 4$ adiabatic expansion
		$4 \rightarrow 1$ isothermal condensation

Figure A.1. Reversible heat engine cycles (Work is performed during path $3 \rightarrow 4$.)

and after the fluid performs work on a piston, it rejects heat Q_C at a lower temperature, T_C. The efficiency of the cycle in converting thermal energy into useful work is obtained by dividing the difference between heat input and rejected heat, by the heat input, i.e.

$$\eta = \frac{Q_H - Q_C}{Q_H} \qquad (A.1)$$

From the expression for the Carnot cycle efficiency, 100% conversion efficiency can only be achieved when $T_C = 0$ K: an impossible situation according to the third law of thermodynamics. In practice, there are other limitations on the attainable efficiency of a heat engine, concerned with the unavoidable losses of energy through inadequate insulation or through friction between moving parts.

On a temperature–entropy (T–S) diagram it is particularly easy to show the relative efficiency of these different cycles. A Carnot cycle consists of two isothermal processes and two adiabatic processes. Now, in a reversible process the heat added, dQ, produces an increase in the entropy of the system, dS, according to the absolute temperature, T, i.e.

$$dQ = T \, dS \qquad (A.2)$$

A *reversible adiabatic* process has no net loss or gain of heat, and so, from equation (A.2), it is also isoentropic. On a T–S diagram this is represented by a vertical line. Thus the Carnot cycle is represented by a rectangle, whose area gives the cycle efficiency. Other cycles under the same operating limits must approach as closely as possible to this area if they are to achieve their maximum efficiency, and it is apparent from figure A.2 that the simple Stirling cycle is better than the simple Rankine cycle in this respect.

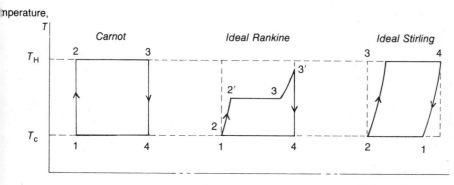

Figure A.2. Comparison of heat engine cycles.

The Rankine cycle may be made more efficient, at the expense of complexity, by superheating the vapour, or by partially expanding it and then reheating it, several times (which show as spikes on the T–S

diagram). Another modification is to pass the liquid through a regenerator on its way to the boiler so that it may be preheated by some of the rejected heat, Q_C. (This is also desirable for the efficient operation of a Stirling cycle.) A further improvement is to use more than one working fluid, so that heat rejected by a high-temperature cycle can be used as input to a cycle using a lower boiling-point fluid, for the greatest efficiency is obtained when most of the energy input is used to vaporize the liquid with the minimum used for raising the liquid to its boiling-point.

The Rankine cycle is complex because it involves a change of state. Consequently, Mollier diagrams of enthalpy–entropy (H–S) changes are employed by engineers as well as the P–V and T–S diagrams used above. Tables of the entropy and enthalpy, with respect to a defined reference, are available for steam and some other substances. To see the significance of this approach we must use the definition of the enthalpy of a system having internal energy, U:

$$H = U + PV \qquad (A.3)$$

Since we are interested in changes of enthalpy around a cycle rather than in absolute values, we can rewrite this equation in terms of the heat added or subtracted:

$$dH = dQ + V\,dP \quad (dQ = dU + P\,dV) \qquad (A.4)$$

(A *reversible adiabatic* process ($dQ = 0$) undergoes a change in enthalpy represented by the area, $V\,dP$, to the left of the adiabatic curve on a P–V diagram.) By combining equations (A.2) and (A.4), it can be seen that the slope of an isobaric process (i.e. $dP = 0$) on a Mollier diagram is the absolute temperature, i.e.

$$\frac{dH}{dS} = T \qquad (A.5)$$

Since heat is added or rejected isobarically in a simple Rankine cycle, these energy transfers are given by the enthalpy change, and the conversion efficiency is readily shown on a Mollier diagram.

Appendix B. Glossary

Absorptance (= **Absorptivity**) a: The ratio of radiant energy absorbed by a surface to that incident on the surface.

Acceptors: Elements from a higher group in the periodic table, added to a semiconductor. These will have one valence electron too few to satisfy the local bonding requirements, and so can accept a free electron.

Adiabatic: An adiabatic change in a system is one in which all three variables, pressure, volume and temperature, may change without heat entering or leaving the system.

Air mass number: The number of thicknesses of atmosphere encountered by the Sun's radiation as it passes to the Earth's surface. AM1 is the shortest possible path through the atmosphere.

Albedo: Reflectance (of the Earth).

Autarkic: Self-sufficient (in energy).

Bandgap (= **Energy gap**) E_g: The range of energy not absorbed by a pure semiconductor or insulator: it is the energy separation of the conduction and valence bands.

Black body: An ideal absorber of radiation, which itself radiates only according to its temperature.

Boltzmann constant, k_B: $1 \cdot 381 \times 10^{-23}$ J K^{-1} or $8 \cdot 617 \times 10^{-5}$ eV K^{-1}.

Campbell–Stokes recorder: An instrument for recording the duration of sunlight, by focusing it through a glass surface on to a paper chart, where it produces a burnt track.

Carnot cycle: The maximum efficiency which can be achieved by a heat-engine operating between a heat source at T_H (K) and a heat sink at T_c(K) is that given by an engine operating on a Carnot cycle:

$$\eta_c = 1 - \frac{T_c}{T_H}$$

Carrier lifetime, τ: Charge carriers injected into a semiconductor will drain away by recombination, with a characteristic time, τ.

Coefficient of performance, CoP: The efficiency of a refrigerator is given by the ratio of the heat extracted to the work performed during the process, or CoP.

Cold mirrors: These have high reflectance for visible wavelength, but transmit in the infra-red.

Concentration ratio, CR: The ratio of the radiant flux on a collector to that on the absorber.

Conduction band, CB: The lowest unoccupied band of electron energy levels in a semiconductor or insulator at 0 K. At higher temperatures it contains some free electrons.

Curve factor, CF: The fill factor of an ideal solar cell, having no resistance losses.

Czochralski growth: The most important method for growing large silicon ingots. A seed crystal is slowly withdrawn from the surface of a crucible of molten silicon, all contained in an inert atmosphere. When under proper control, cylindrical single crystals of several inches diameter and a few feet long can be prepared.

Depletion region: This region in a semiconductor device contains few free carriers, for they are swept out by an electrostatic field produced by fixed ionic charges. It is produced at the junction of any semiconductor with another material.

Diffusion length, L: The average distance that a charge carrier can diffuse in a semiconductor (in the absence of an electric field) before recombining with its complementary partner.

Diffusion potential, V_D: The potential difference across a depletion region, which is produced by free carriers diffusing across a semiconductor junction.

Donors: Elements from a lower group in the periodic table, added to a semiconductor. These will have one valence electron too many for the local bonding requirements, and so can donate a free electron.

Doping: The controlled addition of impurities (usually donors or acceptors) to a semiconductor, to alter its electrical conductivity.

Electron affinity, $e\chi$: In solid-state physics this is the minimum energy required to move an electron from the bottom of the conduction band of a semiconductor into a vacuum. In chemistry it is a measure of the tendency of an atom or molecule to form a negative ion by binding an electron.

Emittance ($=$ emissivity) ε: The ratio of radiant energy emitted by a surface to that emitted by a black body under the same conditions.

Enthalpy, H: The thermodynamic heat content of a system. It is defined by $H = U + PV$, where U is the internal energy of the system.

Entropy, S: Entropy is a measure of the molecular disorder of a system. A *change* in the entropy of a system is defined by the second law of thermodynamics as $dS = dQ/T$, where dQ is the heat *reversibly* added or subtracted, at temperature T. *Irreversible* additions or subtractions of heat occur in the real world, and so the entropy of the Universe is increasing.

Enzyme: A group of proteins synthesized by living cells, which catalyse chemical reactions.

Epitaxial film: One having the same lattice structure as the underlying substrate.

Eutectic: A eutectic mixture of substances is one having a lower freezing point than any of the possible combinations.

Fermi level, E_F: The energy below which electrons exist in a metal or semiconductor at 0 K. At higher temperatures it is the energy level having a 50% probability of being occupied.

Fill factor, FF: The ratio of maximum output power to the product of open-circuit voltage and short-circuit current for a photovoltaic cell.

Heat mirrors: These have high infra-red reflectance and low visible reflectance.

Ideality factor, n: The parameter which enters the exponent in the modified Shockley semiconductor-diode equation:

$$I = I_o \left[\exp(eV/nk_BT) - 1\right]$$

Insolation: The solar irradiance per unit area for a given period.

Internal energy, U: A thermodynamic parameter, defined by reference to changes in its value, like most of these parameters:

$$dU = dQ - dW$$

dW is the work performed on or by a system, with exchange of heat dQ.

Intrinsic carrier concentration, n_i: The density of free carriers (electrons or holes) in an undoped semiconductor. It increases with increase in temperature, and is largest in semiconductors with small bandgaps.

Irradiance: Radiant energy received per unit area per unit time, ($W\ m^{-2}$).

Isothermal: At the same temperature.

Lattice: The regular array of points about which the molecules or ions vibrate in a solid.

Majority and minority carriers: Semiconductors contain both free electrons and holes. In an n-type material the electrons dominate, and are the majority carriers, whilst the holes are the minority carriers (and vice versa for p-type).

n-type: Semiconductor doped with donors, to give a greater net concentration of free electrons than holes.

p-type: Semiconductor doped with acceptors, to give a greater net concentration of free holes than electrons.

Phonon: The quantum of lattice vibrational energy in a solid.

Photochemical effect: A chemical reaction produced by radiation.

Photogalvanic cell: A chemical battery powered by radiation.

Photolysis: A chemical decomposition powered by radiation.

Photovoltaic cell: A semiconductor battery with no associated chemical change, powered by radiation.

Planck constant, h: $6\cdot626 \times 10^{-34}\ J\ s$.

Pyranometer (= solarimeter): An instrument for measuring total solar irradiance: global irradiance is measured if the instrument is horizontal.

Pyrheliometer: An instrument for measuring the direct component of solar irradiance.

Reflectance (= reflectivity) ρ: The ratio of radiant energy reflected from a surface to that incident on the surface.

Reverse saturation current (= leakage current), I_o: The current which flows under reverse bias in a semiconductor diode.

Richardson constant, A: A proportionality factor in the equation which defines the thermionic emission current from a heated metal into vacuum:

$$A = 4\pi emk_B^2/h^3 = 1\cdot2 \times 10^6\ A\ m^{-2}\ K^{-2}$$

For the current flowing across a metal–*semiconductor* junction, a similar equation can be applied, with a Richardson constant modified by the *effective* carrier mass, m^*. (k_B is Boltzmann's constant, and h is Planck's constant.)

Schottky diode: A metal–semiconductor junction diode: in practice most of these devices have an intermediate oxide layer.

Sheet resistivity: The resistance of a square sample of thin-film material, in ohm per square. (All squares have the same resistance whatever their area.)

Sputtering: Vaporization of a solid target, by bombardment with gas ions or atoms in a low-pressure chamber.

Stefan–Boltzmann constant, σ: $5{\cdot}670 \times 10^{-8}\,\mathrm{W\,m^{-2}\,K^{-4}}$.

Superconductivity: As the temperature of metals is reduced towards 0 K, their electrical resistivity disappears and they become superconducting.

Symbiosis: Two organisms co-existing to their mutual benefit.

Thermal expansivity (= coefficient of thermal expansion): The increase in length (or volume, or area) per unit length (volume, area) caused by 1 K temperature rise.

Thermal impedance: A material's resistance to heat flow—the inverse of thermal conductance.

Thermal matching: Ensuring that two materials in contact have compatible thermal expansivity, to avoid stresses building-up during temperature changes.

Transmittance (= Transmissivity) τ: The ratio of radiant energy transmitted to that incident on the sample.

Turbidity coefficient: A factor to allow for the scattering and absorption of solar radiation in the atmosphere by small particles and aerosols.

Valence band VB: The highest occupied band of electron energy levels in a semiconductor or insulator at 0 K. At higher temperatures it will be partially empty—that is, it will contain free holes.

Voltage factor, VF: The ratio of open circuit voltage to bandgap in a photovoltaic cell.

Winston (CPC) collector: A trough-shaped concentrator solar collector with parabolic-section sides, which will transmit radiation to an absorber at the exit only if it enters within a defined angle. *All* radiation entering within this angle will reach the absorber.

Work function, φ_m: The minimum energy required to release an electron at the Fermi level of a metal or semiconductor into vacuum. In the expression $(\varphi/k_B T)$, φ is in joules. It is often expressed in eV, and then stated (for short) in volts.

Appendix C. Energy units conversion chart

The first column gives the wavelength of radiation having the energy given in the other two columns, on the same level.
$1 \text{ eV} = 1 \cdot 24 \, \mu\text{m} = 1 \cdot 60 \times 10^{-19} \text{ J}$.

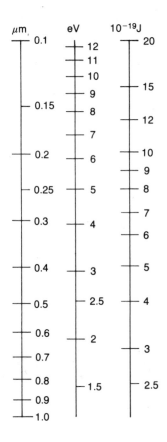

Appendix D. Bibliography

General

UK ISES, May 1976. *Solar Energy—a UK Assessment.*

ISES. *Solar Energy.* Bi-monthly journal.

UK ISES. *Sun at Work in Britain.*

ISES. *Sunworld.* News journal.

A. B. Meinel and M. P. Meinel, 1976. *Applied Solar Energy—An Introduction*, (Reading, Mass., Addison-Wesley). Particularly strong on the optics of solar energy collection.

J. A. Duffie and W. A. Beckman, 1974. *Solar Energy Thermal Processes*, (New York and Chichester, Wiley). The best source for solar thermal collection theory.

A. M. Zarem and D. D. Erway, 1963. *Introduction to the Utilization of Solar Energy*, (New York, McGraw-Hill). One of the classic texts.

Farrington Daniels, 1964. *Direct use of the Sun's Energy*, reprinted by Ballantine Books, New York, 1977. A classic text, covering all aspects of the subject.

Portola Institute, 1974. *Energy Primer: Solar, Water, Wind, and Biofuels*, (Portola Institute). A useful source book for renewable-energy conversion processes.

Proceedings on the annual 'Intersociety Energy Conversion Engineering Conferences', held in USA.

There are several scientific journals which regularly publish articles of interest in the field of energy conversion, e.g. *New Scientist, Scientific American, Science, Mechanical Engineering, Nature, Contemporary Physics.*

A few more specific references are given below.

Chapter 1.
1. World Energy Conference, 1974. *Survey of Energy Resources.*

Chapter 2
1. N. Robinson, 1966. *Solar Radiation*, (Amsterdam, Elsevier).
2. M. P. Thekaekara, 1976. *Solar Energy*, **18,** 309.

Chapter 3
1. F. A. Jenkins and H. E. White, 1976. *Fundamentals of Optics,* (New York, McGraw-Hill).
2. L. A. Azaroff and J. J. Brophy, 1963. *Electronic Processes in Materials*, (New York, McGraw-Hill).

Chapter 6
1. H. P. Maruska and A. K. Ghosh, 1978. *Solar Energy*, **20,** 443.
2. G. Porter and M. D. Archer, 1976. *Interdisciplinary Science Reviews*, **1,** 119.

Chapter 7
1. P. J. Bateman, 1961. *Contemporary Physics*, **2,** 302.
2. C. Backus, ed., 1976. *Solar Cells*, (New York and Chichester, Wiley, IEEE press). A set of reprints of key papers in the field.
3. H. J. Hovel, 1976. *Semiconductors and Semimetals*, Vol. II, *Solar Cells*, edited by A. C. Beer and R. K. Willardson, (New York and London, Academic Press). A briefer version appears in *Solar Energy*, **19,** 605 (1977). The best book for solar cell physics.

Chapter 9
1. J. P. Kemper, 1977. *Sunworld,* **5.**

Chapter 10
1. J. P. McGowan, 1976. *Solar Energy*, **18, 81.**

Chapter 11
1. Progress in solar cell development may be followed in the European and USA solar cell conference proceedings: e.g.
    (*a*) Photovoltaic Solar Energy Conference, Luxembourg, 1977 (Dordrecht, Reidel).
    (*b*) 13th IEEE Photovoltaic Specialists Conference, Washington, D.C., 1978.

Chapter 12
1. J. P. Cooper, 1975. *Photosynthesis and Productivity in Different Environments*, (Cambridge University Press).
2. P. C. Hanawalt and R. H. Haynes, eds., 1973. *The Chemical Basis of Life*, reprinted articles from *Scientific American*, (San Francisco, Freeman).

Chapter 13
1. B. Vale and R. Vale, 1976. *The Autonomous House*, (London, Thames and Hudson).
2. J. A. Sumner, 1976. *Domestic Heat Pumps*, (Prism Press).
3. W. Palz, 1978. *Solar Electricity—An Economic Approach to Solar Energy* (UNESCO/Butterworths).
4. P. E. Glaser, 1977. *Physics Today,* **30,** 30.

Chapter 14
1. *CEGB Research*, 1975, **2** (May).

Index

Entries in italics refer to definitions in the glossary